CW00548711

The Origins of Na

The Origins of Natural Science

Nine lectures delivered in Dornach
December 24 to 28, 1922 and January 1 to 6, 1923

by
Rudolf Steiner

with an Introduction by
Owen Barfield

Anthroposophic Press
Spring Valley, NY

Rudolf Steiner Press
London

The nine lectures presented here were given in Dornach from Dec. 24 to 28, 1922 and from Jan. 1 to 6, 1923. In the Collected Edition of Rudolf Steiner's works, the volume containing the German texts is entitled, *Der Entstehungsmoment der Naturwissenschaft in der Weltgeschichte und ihre seitherige Entwickelung.* (Vol. 326 in the Collected Edition.) The lectures were translated by Maria St. Goar and edited by Norman Macbeth.

Copyright © 1985
Anthroposophic Press

Library of Congress Cataloging in Publication Data

Steiner, Rudolf, 1861–1925.
     The origins of natural science.

     Translation of: Die Entstehungsmoment der Naturwissenschaft in der Weltgeschichte and ihre seitherige Entwickelung.
     1. Anthroposophy—Addresses. essays. lectures.
2. Science—Addresses, essays, lectures. I. Macbeth, Norman, 1910–   . II. Title.
BP595.S894E5713 1985   299'.935   85-11197
ISBN 0-88010-140-5
ISBN 0-88010-140-7 (pbk.)

Cover design by Peter Stebbing

All rights reserved. No part of this book may be reproduced in any form without the written permission of the publishers, except for brief quotations embodied in critical articles and reviews.

Published in the US by Anthroposophic Press, 258 Hungry Hollow Rd, Spring Valley, NY 10977
Published in the UK by Rudolf Steiner Press, 38 Museum St., London WC 1

Table of Contents

Introduction

The nine lectures that follow were delivered by Rudolf Steiner at the turn of 1922/23 in Dornach, Switzerland. They were directed to an audience containing some professional scientists and others particularly interested in science, many of whom were members of the Anthroposophical Society. 1922/23 happens also to have been an historical moment in the life of the Society and indeed of the lecturer. No one reading them would suspect that between Lectures V and VI both parties had been stricken by a crushing blow. On New Year's Eve 1922 the building named the Goetheanum, in which the first five lectures had been given, was totally destroyed by fire and was indeed still burning on January 1 when Steiner delivered Lecture VI in his private Studio. The great wooden structure, a temple rather than a mere headquarters or meeting-place, had been designed by Steiner himself, its building supervised by him at all stages, and much of its interior worked with his own hands; but this is not the place to enlarge on his personal tragedy or the courage and determination it must have required to continue with the lecture course on the following day almost as if nothing had happened. The most that critical appraisal might detect as a possible consequence of that grievous interruption is perhaps a certain repetitiveness not apparent elsewhere in either his books or his lectures; and this the translator has taken the liberty of slightly reducing.

One more preliminary observation may be desirable. Most members of the original audience would have been familiar, to a greater or less degree, with the fundamental teachings and thus with the terminology of anthroposophy, or spiritual science, as Steiner also named them. Here and there in the lectures some of that terminology is introduced, for example "etheric," "astral," "the Age of the consciousness soul." Mostly their meaning is briefly indicated when they first appear; but it remains true that some previous familiarity with them is of considerable assistance towards a full understanding, not only of particular passages, but also of the radical message of the whole.

Their basic argument is that modern science, and the scientism based on it, so far from being the only possible 'reality-principle' is merely one way of conceiving the nature of reality; a way moreover that has arisen only recently and which there is no reason to suppose will last forever. Many today might admit as much, but in doing so they would be thinking of modern science mainly as a theory or set of theories capable of proof or disproof by accepted methods. For Steiner modern science, *including* its empirical method, is a stage, and an important stage, in the whole evolution of human consciousness. And that is something different from, though it underlies, the history of ideas. Perception itself is determined by the human psyche, the consciousness which determines perception precedes the formation of thoughts based on that perception, and the human psyche is an evolving one. Only hitherto it has not been conscious of that fact. Certain ideas were formed, and could only be formed, at certain stages in that evolution. Ideas for instance or theories about the nature of the world, or the nature of Nature, are necessarily based on certain 'givens'—experiences taken for granted—which are so immediate that no ideas at all can be formed about *them*. Isaac Newton, as Lecture III points out, was sufficiently aware of this to declare the 'givens' of his own day as the 'postulates' from which he started. They were time, place, space and motion. And these remain the givens for our day, even if their slight unsettling by Einstein's relativity should be the first faint breath of coming winds of change. But they were not so for other days and other men. They were not so before at most the fifteenth century. They are given for us, because for us the outer world of natural objects and events is experienced as completely detached from the inner world of our own awareness of them, that is to say, from our humanity. Descartes was the first to formulate this—then comparatively novel—given, when he divided the world into extended substance and thinking substance.

Writing in 1818 an essay on *Method*, Coleridge prophesied:

> . . . there will soon be seen a general tendency toward, an earnest seeking after, some ground common to the world and to man, therein to find the one principle of permanence and identity, the rock of strength and refuge, to which the soul may cling amid the fleeting surge-like objects of the sense.

The abiding thrust of these lectures is Steiner's unshakeable conviction that from now on the progress of science will depend on the overcoming of the received dichotomy between man and nature just as from the fifteenth or sixteenth century *up* to now the progress of science has depended *on* that dichotomy. Incidental to that progress would be escape from the crudities of popular scientism, but the lectures are only marginally concerned with that. Their content is based on the fact that the understanding, perhaps of any phenomenon but certainly of any phenomenon so basic as to be 'given,' entails a patient examination of its provenance, that is to say of the steps by which it came into being. Consequently they are, as the title suggests, lectures not on science, but on the history of science. In sum they tell the story of the origin and then of the growth of that gulf between *inner* and *outer*, between subject and object, extending from a time before Pythagoras down to our own day, as it is manifest in the writings and biographies of a selection of well-known thinkers. Particular attention is given to transitional figures, men whose perceptions were still determined by the past, while their thoughts were confronted by what was approaching from the future; and perhaps especially interesting in this regard are the observations on Giordano Bruno's cosmos in Lecture IV and Galen's theory of 'fermentation' in Lecture VIII.

The story is at the same time one of the steadily increasing predominence of mathematics in determining scientific method. Perception of this is not peculiar to Steiner. What distinguishes him from other historians of science is the psychological detail into which he pursues the story and, more than that, his account of the *origin* of mathematics. The Cartesian coordinates are not as abstract as they seem; or rather they were not always so. Steiner sees them as an extrapolation or projection of man's experience of his own body; that is to say, of his physical body. And here is one of the places where some previous acquaintance with anthroposophy and its terminology would be helpful, though it should not be indispensable. It is unfortunate that the word "body" has become, for most people, almost synonymous with "lump of solid matter"; particularly unfortunate, where it is the human body that is at issue, since nine-tenths of that is composed of fluids, and of fluids that are for the most part in motion. "Body" in Steiner's terminology, signifies something more like "systematically organised unit or entity," as distinct from the matter or substance of which it is composed. Thus, the fact that the frame of a living human being contains, and not at random, fluid and airy, as well as solid, substance, entails the existence of other "bodies" besides the physically organised one. These are especially relevant when the discourse turns from knowledge of quantity (measurement and mathematics) to knowledge of quality, an aspect of nature that is virtually a closed book to the science of today.

The development of that science of today, a purely quantitative one, is the main thread on which the lectures are strung, and the reader will follow it for himself. Not much perhaps would be gained by informing him in advance that, if he does so, he will be shown for example, how the projection of mathematics, and particularly the coordinates, outward from the body and thus from human selfhood, has led to the reification of space—that long-settled mental habit which advanced physics has only recently begun to question. He will also find an answer to a question which has puzzled many thinkers: *why* should mathematics, a seemingly artificial construction of the human brain, have been found an effective key to unlock so many of the secrets of nature? How is it that the one has happened to fit so snugly on the other? More generally he will be led down a sort of ladder of "descent," accompanied throughout by mathematics, from man's original psychic participation in the life of nature to his present detachment from it; to be shown at the end that an understanding of the way of ascent to reunion with that life also *begins* with mathematics. The last is an aspect of the matter with which Steiner was to deal more specifically in a subsequent course of lectures translated into English as *The Boundaries of Natural Science.*

"Descent" and "ascent" are of course loaded terms, and their use can be misleading. The same is true of the term "dehumanization" when in these lectures it is applied to the history of science. Steiner was no enemy of science, though he vigorously questioned many of its theories. "Technology" is not a dirty word in his vocabulary. Pointing to a fact is not necessarily abuse. Science has become dehumanized in the sense that it has turned its attention more and more away from human experience and human values. But in doing so it has furthered, if not partly engendered, one supreme human value—that detached, individual self-consciousness that is the pre-condition of freedom. Man has become separated from the world that gave him birth; but he needed that separation in order to become truly man. To draw attention to that separation is, says our lecturer, "a description of the scientific view, not a criticism." He continues (and I will conclude this Introduction by quoting the closing words of Lecture VI):

> Let us assume that somebody says: "Here I have water. I cannot use it in this state. I separate the oxygen from the hydrogen, because I need the hydrogen." He then proceeds to do so. If I then say what he has done, this is not criticism of his conduct. I have no business to tell him he is doing something wrong, and should leave the water alone. Nor is it criticism when I say that since the Fifteenth Century science has taken the world of living beings and separated it from the true nature of man, discarding it and retaining what this age required. It then led this dehumanized science to the triumphs that have been achieved.
>
> It is not criticism if something like this is said: it is only a description. The scientist of modern times needed a dehumanized nature, just as a chemist needs deoxygenized hydrogen and therefore has to split water into its two components. The point is to understand that we must not constantly fall into the error of looking to science for an understanding of man.

LECTURE I

Dornach, December 24, 1922

My dear friends! You have come together this Christmas, some of you from distant places, to work in the Goetheanum on some matters in the field of spiritual science. At the outset of our considerations I would like to extend to you—especially the friends who have come from afar—our heartiest Christmas greetings. What I myself, occupied as I am with the most manifold tasks, will be able to offer you at this particular time can only be indications in one or another direction. Such indications as will be offered in my lectures, and in those of others, will, we hope, result in a harmony of feeling and thinking among those gathered together here in the Goetheanum. It is also my hope that those friends who are associated with the Goetheanum and more or less permanently residing here will warmly welcome those who have come from elsewhere. Through our working, thinking and feeling together, there will develop what must be the very soul of all endeavors at the Goetheanum; namely, our perceiving and working out of the spiritual life and essence of the world.

If this ideal increasingly becomes a reality, if the efforts of individuals interested in the anthroposophical world conception flow together in true social cooperation, in mutual give and take, then there will emerge what is intended to emerge at the Goetheanum. In this spirit, I extend the heartiest welcome to those friends who have come here from afar as well as to those residing more permanently in Dornach.

The indications that I shall try to give in this lecture

1

course will not at first sight appear to be related to the thought and feeling of Christmas, yet inwardly, I believe, they are so related. In all that is to be achieved at the Goetheanum, we are striving toward the birth of something new, toward knowledge of the spirit, toward a feeling consecrated to the spirit, toward a will sustained by the spirit. This is in a sense the birth of a supersensible spiritual element and, in a very real way, symbolizes the Christmas thought, the birth of that spiritual Being who produced a renewal of all human evolution upon earth. Therefore, our present studies are, after all, imbued with the character of a Christmas study.

Our aim in these lectures is to establish the moment in history when the scientific mode of thinking entered mankind's development. This does not conflict with what I have just said. If you remember what I described many years ago in my book *Mysticism at the Dawn of Modern Age*[1], you will perceive my conviction that beneath the external trappings of scientific conceptions one can see the first beginnings of a new spirituality. My opinion, based on objective study, is that the scientific path taken by modern humanity was, if rightly understood, not erroneous but entirely proper. Moreover, if regarded in the right way, it bears within itself the seed of a new perception and a new spiritual activity of will. It is from this point of view that I would like to give these lectures. They will not aim at any kind of opposition to science. The aim and intent is instead to discover the seeds of spiritual life in the highly productive modern methods of scientific research. On many occasions I have pointed this out in various ways. In lectures given at various times on various areas of natural scientific thinking,[2] I have given details of the path that I want to characterize in broader outline during the present lectures.

If we want to acquaint ourselves with the real meaning of scientific research in recent times and the mode of thinking that can and does underlie it, we must go back several centuries into the past. The essence of scientific thinking is easily misunderstood, if we look only at the immediate present. The actual nature of scientific research cannot be understood unless its development is traced through several centuries. We must go back to a point in time that I have often described as very significant in modern evolution; namely, the fourteenth and fifteenth centuries. At that time, an altogether different form of thinking, which was still active through the Middle Ages, was supplanted by the dawn of our present-day mode of thought. As we look back into this dawn of the modern age, in which many memories of the past were still alive, we encounter a man in whom we can see, as it were, the whole transition from an earlier to a later form of thinking. He is Cardinal Nicholas Cusanus,[3] a renowned churchman and one of the greatest thinkers of all time. He was born in 1401, the son of a boatman and vine-grower in the Rhine country of Western Germany, and died in 1464, a persecuted ecclesiastic.[4] Though he may have understood himself quite well, Cusanus was a person who is in some respects extremely difficult for a modern student to comprehend.

Cusanus received his early education in the community that has been called ''The Brethren of the Common Life.''[5] There he absorbed his earliest impressions, which were of a peculiar kind. It is clear that Nicholas already possessed a certain amount of ambition as a boy, but this was tempered by an extraordinary gift for comprehending the needs of the social life of his time. In the community of the Brethren of the Common Life, persons were gathered together who were dissatisfied with the church institutions and with the monastic and religious orders that, though within the church, were to some degree in opposition to it.

In a manner of speaking, the Brethren of the Common Life were mystical revolutionaries. They wanted to attain

3

what they regarded as their ideal purely by intensification of a life spent in peace and human brotherhood. They rejected any rulership based on power, such as was found in a most objectionable form in the official church of that time. They did not want to become estranged from the world as were members of monastic orders. They stressed physical cleanliness; they insisted that each one should faithfully and diligently perform his duty in external life and in his profession. They did not want to withdraw from the world. In a life devoted to genuine work they only wanted to withdraw from time to time into the depths of their souls. Alongside the external reality of life, which they acknowledged fully in a practical sense, they wanted to discover the depths and inwardness of religious and spiritual feeling. Theirs was a community that above all else cultivated human qualities in an atmosphere where a certain intimacy with God and contemplation of the spirit might abide. It was in this community—at Deventer in Holland—that Cusanus was educated. The majority of the members were people who, in rather narrow circles, fulfilled their duties, and sought in their quiet chambers for God and the spiritual world.

Cusanus, on the other hand, was by nature disposed to be active in outer life and, through the strength of will springing from his knowledge, to involve himself in organizing social life. Thus Cusanus soon felt impelled to leave the intimacy of life in the brotherhood and enter the outer world. At first, he accomplished this by studying jurisprudence. It must be borne in mind, however, that at that time—the early Fifteenth Century—the various sciences were less specialized and had many more points of contact than was the case later on.

So for a while Cusanus practiced law. His was an era, however, in which chaotic factors extended into all spheres of social life. He therefore soon wearied of his law practice and had himself ordained a priest of the Roman Catholic

Church. He always put his whole heart into whatever he did, and so he now became a true priest of the Papal church. He worked in this capacity in the various clerical posts assigned to him, and he was particularly active at the Council of Basle (1431-1449).[6] There he headed a minority whose ultimate aim it was to uphold the absolute power of the Holy See.[7] The majority, consisting for the most part of bishops and cardinals from the West, were striving after a more democratic form, so to speak, of church administration. The pope, they thought, should be subordinated to the councils. This led to a schism in the Council. Those who followed Cusanus moved the seat of the Council to the South; the others remained in Basle and set up an anti-pope.[8] Cusanus remained firm in his defense of an absolute papacy. With a little insight it is easy to imagine the feelings that impelled Cusanus to take this stand. He must have felt that whatever emerged from a majority could at best lead only to a somewhat sublimated form of the same chaos already existing in his day. What he wanted was a firm hand that would bring about law and order, though he did want firmness permeated with insight. When he was sent to Middle Europe later on, he made good this desire by upholding consolidation of the Papal church.[9] He was therefore, as a matter of course, destined to become a cardinal of the Papal church of that time.

As I said earlier, Nicholas probably understood himself quite well, but a latter-day observer finds him hard to understand. This becomes particularly evident when we see this defender of absolute papal power traveling from place to place and—if the words he then spoke are taken at face value—fanatically upholding the papistical Christianity of the West against the impending danger of a Turkish invasion.[10] On the one hand, Cusanus (who in all likelihood had already been made a cardinal by that time) spoke in flaming words against the infidels. In vehement terms he summoned

Europe to unite in resistance to the Turkish threat from Asia. On the other hand, if we study a book that Cusanus probably composed[11] in the very midst of his inflammatory campaigns against the Turks, we find something strange. In the first place, Cusanus preaches in the most rousing manner against the imminent danger posed by the Turks, inciting all good men to defend themselves against this peril and thus save European civilization. But then Cusanus sits down at his desk and writes a treatise on how Christians and Jews, pagans and Moslems—provided they are rightly understood —can all be brought to peaceful cooperation, to the worship and recognition of the one universal God; how in Christians, Jews, Moslems and heathens there dwells a common element that need only be discovered to create peace among mankind. Thus the most conciliatory sentiments in regard to religions and denominations flow from this man's quiet private chamber, while he publicly calls for war in the most fanatical words.

This is what makes it hard to understand a man like Nicholas Cusanus. Only real insight into that age can make him comprehensible but he must be viewed in the context of the inner spiritual development of his time. No criticism is intended. We only want to see the external side of this man, with the furious activity that I have described, and then to see what was living in his soul. We simply want to place the two aspects side by side.

We can best observe what took place in Cusanus's mind if we study the mood he was in while returning from a mission to Constantinople[12] on behalf of the Holy See. His task was to work for the reconcilation of the Western and Eastern churches. On his return voyage, when he was on the ship and looking at the stars, there arose in him the fundamental thought, the basic feeling, incorporated in the book that he published in 1440 under the title *De Docta Ignorantia (On Learned Ignorance)*.[13]

What is the mood of this book? Cardinal Cusanus had, of course, long since absorbed all the spiritual knowledge current in the Middle Ages. He was well versed also in what the medieval schools of Neo-Platonism and Neo-Aristotelianism had attained. He was also quite familiar with the way Thomas Aquinas had spoken of the spiritual worlds as though it were the most normal thing for human concepts to rise from sense perception to spirit perception. In addition to his mastery of medieval theology, he had a thorough knowledge of the mathematical conceptions accessible to men of that time. He was an exceptionally good mathematician. His soul, therefore, was filled on the one side with the desire to rise through theological concepts to the world of spirit that reveals itself to man as the divine and, on the other side, with all the inner discipline, rigor, and confidence that come to a man who immerses himself in mathematics. Thus he was both a fervent and an accurate thinker.

When he was crossing the sea from Constantinople to the West and looking up at the starlit sky, his twofold soul mood characterized above resolved itself in the following feeling. Thenceforth, Cusanus conceived the deity as something lying outside human knowledge. He told himself: "We can live here on earth with our knowledge, with our concepts and thoughts. By means of these we can take hold of what surrounds us in the kingdoms of nature. But these concepts grow ever more lame when we direct our gaze upward to what reveals itself as the divine."

In Scholasticism, arising from quite another viewpoint, a gap had opened up between knowledge and revelation.[14] This gap now became the deepest problem of Cusanus's soul, the most intimate concern of the heart. Repeatedly he went through this course of reasoning, repeatedly he saw how thinking extends itself over everything surrounding man in nature; how it then tries to raise itself above this realm to the divinity of thoughts; and how, there, it

7

becomes ever more tenuous until it finally completely dissipates into nothingness as it realizes that the divine lies beyond that void into which thinking has dissipated. Only if a man has developed (apart from his life in thought) sufficient fervent love to be capable of continuing further on this path that his thought has traversed, only if love gains the lead over thought, then this love can attain the realm into which knowledge gained only by thinking cannot reach.

It therefore became a matter of deep concern for Cusanus to designate the actual divine realm as the dimension before which human thought grows lame and human knowledge is dispersed into nothingness. This was his *docta ignorantia*, his learned ignorance. Nicholas Cusanus felt that when erudition, knowledge, assumes in the noblest sense a state of renouncing itself at the instant when it thinks to attain the spirit, then it achieves its highest form, it becomes *docta ignorantia*. It was in this mood that Cusanus published his *De Docta Ignorantia* in 1440.

Let us leave Cusanus for the moment, and look into the lonely cell of a medieval mystic who preceded Cusanus. To the extent that this man has significance for spiritual science, I described him in my book on mysticism. He is Meister Eckhart,[15] a man who was declared a heretic by the official church. There are many ways to study the writings of Meister Eckhart and one can delight in the fervor of his mysticism. It is perhaps most profoundly touching if, through repeated study, the reader comes upon a fundamental mood of Eckhart's soul. I would like to describe it as follows. Though living earlier than Cusanus, Meister Eckhart too was imbued through and through with what medieval Christian theology sought as an ascent to the divine, to the spiritual world. When we study Meister Eckhart's writings, we can recognize Thomistic shades of thought in many of his lines. But each time Meister Eckhart's soul tries to rise from theological thinking to the actual spiritual world (with

8

which it feels united), it ends by saying to itself that with all this thinking and theology it cannot penetrate to its innermost essence, to the divine inner spark. It tells itself: This thinking, this theology, these ideas, give me fragments of something here, there, everywhere. But none of these are anything like the spiritual divine spark in my own inner being. Therefore, I am excluded from all thoughts, feelings, and memories that fill my soul, from all knowledge of the world that I can absorb up to the highest level. I am excluded from it all, even though I am seeking the deepest nature of my own being. I am in nothingness when I seek this essence of myself. I have searched and searched. I traveled many paths, and they brought me many ideas and feelings, and on these paths I found much. I searched for my "I," but before ever I found it, I fell into "nothingness" in this search for the "I," although all the kingdoms of nature urged me to the search.

So, in his search for the self, Meister Eckhart felt that he had fallen into nothingness. This feeling evoked in this medieval mystic words that profoundly touch the heart and soul. They can be paraphrased thus: "I submerge myself in God's nothingness, and am eternally, through nothingness, through nothing, an I; through nothing, I become an I. In all eternity, I must fetch the I from the "nothingness" of God."[16] These are powerful words. Why did this urge for "nothing," for finding that I in nothingness, resound in the innermost chamber of this mystic's heart, when he wanted to pass from seeking the world to seeking the I? Why? If we go back into earlier times, we find that in former ages it was possible, when the soul turned its gaze inward into itself, to behold the spirit shining forth within. This was still a heritage of primeval pneumatology, of which we shall speak later on. When Thomas Aquinas, for example, peered into the soul, he found within the soul a weaving, living spiritual element. Thomas Aquinas[17] and his predecessors sought

9

the essential ego not in the soul itself but in the spiritual dwelling in the soul. They looked through the soul into the spirit, and in the spirit they found their God-given I. And they said, or could have said: I penetrate into my inmost soul, gaze into the spirit, and in the spirit I find the I.—In the meantime, however, in humanity's forward development toward the realm of freedom, men had lost the ability to find the spirit when they looked inward into themselves.

An earlier figure such as John Scotus Erigena (810-880 A.D.) would not have spoken as did Meister Eckhart. He would have said: I gaze into my being. When I have traversed all the paths that led me through the kingdoms of the outer world, then I discover the spirit in my inmost soul. Thereby, I find the "I" weaving and living in the soul. I sink myself as spirit into the Divine and discover "I."

It was, alas, human destiny that the path that was still accessible to mankind in earlier centuries was no longer open in Meister Eckhart's time. Exploring along the same avenues as John Scotus Erigena or even Thomas Aquinas, Meister Eckhart could not sink himself into God-the-Spirit, but only into the "nothingness" of the Divine, and from this "nothing" he had to take hold of the I. This shows that mankind could no longer see the spirit in inner vision. Meister Eckhart brought the I out of the naught through the deep fervor of his heart. His successor, Nicholas Cusanus,[18] admits with complete candor: All thoughts and ideas that lead us in our exploration of the world become lame, become as nothing, when we would venture into the realm of spirit. The soul has lost the power to find the spirit realm in its inner being. So Cusanus says to himself: When I experience everything that theology can give me, I am led into this naught of human thinking. I must unite myself with what dwells in this nothingness in order to at least gain in the *docta ignorantia* the experience of the spirit.—Then, however, such knowledge, such perception, cannot be expressed in words. Man is rendered dumb when he has reached the point at which he can experience the spirit only through the *docta ignorantia.*

Thus Cusanus is the man who in his own personal development experiences the end of medieval theology and is driven to the *docta ignorantia.* He is, however, at the same time a skillful mathematician. He has the disciplined thinking that derives from the pursuit of mathematics. But he shies away, as it were, from applying his mathematical skills to the *docta ignorantia.* He approaches the *docta ignorantia* with all kinds of mathematical symbols and formulas, but he does this timidly, diffidently. He is always conscious of the fact that these are symbols derived from mathematics. He says to himself: Mathematics is the last remnant left to me from ancient knowledge. I cannot doubt its reliability as I can doubt that of theology, because I actually experience its reliability when I apprehend mathematics with my mind.—At the same time, his disappointment with theology is so great he dares not apply his mathematical skills in the field of the *docta ignorantia* except in the form of symbols.

This is the end of one epoch in human thinking. In his inner mood of soul, Cusanus was almost as much of a mathematician as was Descartes later on, but he dared not try to grasp with mathematics what appeared to him in the manner he described in his *Docta Ignorantia.* He felt as though the spirit realm had withdrawn from mankind, had vanished increasingly into the distance, and was unattainable with human knowledge. Man must become ignorant in the innermost sense in order to unite himself in love with this realm of the spirit.

This mood pervades Cusanus's *Docta Ignorantia* published in 1440. In the development of Western civilization, men had once believed that they confronted the spirit-realm in close perspective. But then, this spirit realm became more and more remote from those men who observed it, and finally it vanished. The book of 1440 was a frank admission that the ordinary human comprehension of that time could no longer reach the remote perspectives into which the spirit realm has withdrawn. Mathematics, the most reliable of the sciences, dared to approach only with symbolic formulas what was no longer beheld by the soul. It was as though this spirit realm, receding further and further in perspective, had disappeared from European civilization. But from the opposite direction, another realm was coming increasingly into view. This was the realm of the sense world, which European civilization was beginning to observe and like. In 1440, Nicholas Cusanus applied mathematical thinking and mathematical knowledge to the vanishing spirit realm only by a timid use of symbols; but now Nicholas Copernicus boldly and firmly applied them to the outer sense world. In 1440 the *Docta Ignorantia* appeared with the admission that even with mathematics one can no longer behold the spirit realm. We must conceive the spirit realm as so far removed from human perception that even mathematics can approach it only with halting symbols; this is what Nicholas Cusanus said in 1440. "Conceive of mathematics as so powerful and reliable that it can force the sense world into mathematical formulas that are scientifically understandable." This is what Nicholas Copernicus said to European civilization in 1543. In 1543 Copernicus published his *De Revolutionibus Orbium Coelestium (On the Revolutions of the Celestial Bodies)*, where the universe was depicted so boldly and rudely that it had to surrender itself to mathematical treatment.

One century lies between the two. During this century Western science was born. Earlier, it had been in an embryonic state. Whoever wants to understand what led to the birth of Western science, must understand this century that lies between the *Docta Ignorantia* and the *De Revolutionibus Orbium Coelestium*. Even today, if we are to understand the true meaning of science, we must study the fructifications that occurred at that time in human soul life and the renunciations it had to experience. We must go back this far in time. If we want to have the right scientific attitude, we must begin there, and we must also briefly consider the embryonic state preceding Nicholas Cusanus. Only then can we really comprehend what science can accomplish for mankind and see how new spiritual life can blossom forth from it.

LECTURE II

Dornach, December 25, 1922

The view of history forming the basis of these lectures may be called symptomatological. What takes place in the depths of human evolution sends out waves, and these waves are the symptoms that we will try to describe and interpret. In any serious study of history, this must be the case. The processes and events occurring at any given time in the depths of evolution are so manifold and so significant that we can never do more than hint at what is going on in the depths. This we do by describing the waves that are flung up. They are symptoms of what is actually taking place.

I mention this because, in order to characterize the birth of the scientific form of thinking and research I described two men, Meister Eckhart and Nicholas Cusanus, in my last lecture. What can be historically observed in the soul and appearance of such men I consider to be symptoms of what goes on in the depths of general human development; this is why I give such descriptions. There are in any given case only a couple of images cast up to the surface that we can intercept by looking into one or another soul. Yet, by doing this, we can describe the basic nature of successive time periods.

When I described Cusanus yesterday, my intention was to suggest how all that happened in the early fifteenth century in mankind's spiritual development, which was pressing forward to the scientific method of perception, is symptomatically revealed in his soul. Neither the knowledge that the mind can gather through the study of theology nor the precise perceptions of mathematics can lead any longer to a grasp of the spiritual world. The wealth of human knowledge, its concepts and ideas, come to a halt before that realm. The fact that one can do no more than write a "docta ignorantia" in the face of the spiritual world comes to expression in Cusanus in a remarkable way. He could go no further with the form of knowledge that, up to his time, was prevalent in human development.

As I pointed out, this soul mood was already present in Meister Eckhart. He was well versed in medieval theological knowledge. With it, he attempted to look into his own soul and to find therein the way to the divine spiritual foundations. Meister Eckhart arrived at a soul mood that I illustrated with one of his sentences. He said—and he made many similar statements—"I sink myself into the naught of the divine, and out of nothing become an I in eternity." He felt himself arriving at nothingness with traditional knowledge. Out of this nothingness, after the ancient wisdom's loss of all persuasive power he had to produce out of his own soul the assurance of his own I, and he did it by this statement.

Looking into this matter more closely, we see how a man like Meister Eckhart points to an older knowledge that has come down to him through the course of evolution. It is knowledge that still gave man something of which he could say: This lives in me, it is something divine in me, it *is* something.—But now, in Meister Eckhart's own time, the most profound thinkers had been reduced to the admission: When I seek this something here or there, all knowledge of this something does not suffice to bring me certainty of my own being. I must proceed from the Something to the Nothing and then, in an act of creation, kindle to life the consciousness of self out of naught.

Now, I want to place another man over against these two. This other man lived 2,000 years earlier and for his time he was as characteristic as Cusanus (who followed in Meister Eckhart's footsteps) was for the fifteenth century. This backward glance into ancient times is necessary so that we can better understand the quest for knowledge that surfaced in the Fifteenth Century from the depths of the human soul. The man whom I want to speak about today is not mentioned in any history book or historical document, for these do not go back as far as the Eighth Century B.C. Yet, we can only gain information concerning the origin of science if, through spiritual science, through purely spiritual observation, we go farther back than external historical documents can take us. The man I have in mind lived about 2,000 years prior to the present period (the starting point of which I have assigned to the first half of the fifteenth century). This man of pre-Christian times was accepted into a so-called mystery school of Southeastern Europe. There he heard everything that the teachers of the mysteries could communicate to their pupils concerning spiritual wisdom, truths concerning the spiritual beings that lived and still live in the cosmos. But the wisdom that this man received from his teachers was already more or less traditional. It was a recollection of far older visions, a recapitulation of what wise men of a much more ancient age had beheld when they directed their clairvoyant sight into the cosmic spaces whence the motions and constellations of the stars had spoken to them. To the sages of old, the universe was not the machine, the mechanical contraption that it is for men of today when they look out into space to the wise men of ancient times. The cosmic spaces were like living beings, permeating everything with spirit and speaking to them in cosmic language. They experienced themselves within the spirit of world being. They felt how this, in which they lived and moved, spoke to them, how they could direct their questions concerning the riddles of the universe to the universe itself and how, out of the widths of space, the cosmic

16

phenomena replied to them. This is how they experienced what we, in a weak and abstract way, call "spirit" in our language. Spirit was experienced as the element that is everywhere and can be perceived from anywhere. Men perceived things that even the Greeks no longer beheld with the eye of the soul, things that had faded into a nothingness for the Greeks.

This nothingness of the Greeks, which had been filled with living content for the earliest wise men of the Post-Atlantean age,[19] was named by means of words customary for that time. Translated into our language, though weakened and abstract, those words would signify "spirit." What later became the unknown, the hidden god, was called spirit in those ages when he was known. This is the first thing to know about those ancient times.

The second thing to know is that when a man looked with his soul and spirit vision into himself, he beheld his soul. He experienced it as originating from the spirit that later on became the unknown god. The experience of the ancient sage was such that he designated the human soul with a term that would translate in our language into "spirit messenger" or simply "messenger."

If we put into a diagram what was actually seen in those earliest times, we can say: The spirit was considered the world-embracing element, apart from which there was nothing and by which everything was permeated. This spirit, which was directly perceptible in its archetypal form, was sought and found in the human soul, inasmuch as the latter recognized itself as the messenger of this spirit. Thus the soul was referred to as the "messenger."

A third aspect was external nature with all that today is called the world of physical matter, of bodies. I said above that apart from spirit there was nothing, because spirit was perceived by direct vision everywhere in its archetypal form. It was seen in the soul, which realized the spirit's

17

```
Spirit:  Archetypal Form        S p i r i t
Soul:    Messenger                  /    \
Body:    Image                    /        \
                               Soul        Physical World
                               Messenger   Image
```

message in its own life. But the spirit was likewise recognized in what we call nature today, the world of corporeal things. Even this bodily world was looked upon as an image of the spirit.

In those ancient times, people did not have the conceptions that we have today of the physical world. Wherever they looked, at whatever thing or form of nature, they beheld an image of the spirit, because they were still capable of seeing the spirit. The image nearest to man was the human body, a fragment of nature. Inasmuch as all other phenomena of nature were images of the spirit, the body of man too was an image of the spirit. So when this ancient man looked at himself, he recognized himself as a threefold being. In the first place, the spirit lived in him as in one of its many mansions. Man knew himself as spirit. Secondly, man experienced himself within the world as a messenger of this spirit, hence as a soul being. Thirdly, man experienced his corporeality; and by means of this body he felt himself to be an image of the spirit.[20] Hence, when man looked upon his own being, he perceived himself as a threefold entity of spirit, soul, and body: as spirit in his archtypal form; as soul, the messenger of god; as body, the image of the spirit.

This ancient wisdom contained no contradiction between body and soul or between nature and spirit, because one knew: Spirit is in man in its archetypal form; the soul is none other than the message transmitted by spirit; the body is the image of spirit. Likewise, no contrast was felt between man and surrounding nature because one bore an image of

18

spirit in one's own body, and the same was true of every body in external nature. Hence, an inner kinship was experienced between one's own body and those in outer nature, and nature was not felt to be different from oneself. Man felt himself at one with the whole world. He could feel this because he could behold the archetype of spirit and because the cosmic expanses spoke to him. In consequence of the universe speaking to man, science simply could not exist. Just as we today cannot build a science of external nature out of what lives in our memory, ancient man could not develop one because, whether he looked into himself or outward at nature, he beheld the same image of spirit. No contrast existed between man himself and nature, and there was none between soul and body. The correspondence of soul and body was such that, in a manner of speaking, the body was only the vessel, the artistic reproduction, of the spiritual archetype, while the soul was the mediating messenger between the two. Everything was in a state of intimate union. There could be no question of comprehending anything. We grasp and comprehend what is outside our own life. Anything that we carry within ourselves is directly experienced and need not be first comprehended.

Prior to Roman and Greek times, this wisdom born of direct perception still lived in the mysteries. The man I referred to above heard about this wisdom, but he realized that the teachers in his mystery school were speaking to him only out of a tradition preserved from earlier ages. He no longer heard anything original, anything gained by listening to the secrets of the cosmos. This man undertook long journeys and visited other mystery centers, but it was the same wherever he went. Already in the Eighth Century B.C., only traditions of the ancient wisdom were preserved everywhere. The pupils learned them from the teachers, but the teachers could no longer see them, at least not in the vividness of ancient times.

But this man whom I have in mind had an unappeasable urge for certainty and knowledge. From the communications passed on to him, he gathered that once upon a time men had indeed been able to hear the harmony of the spheres from which resounded the Logos that was identical with the spiritual archetype of all things. Now, however, it was all tradition. Just as 2,000 years later Meister Eckhart, working out the traditions of his age, withdrew into his quiet monastic cell in search of the inner power source of soul and self, and at length came to say: "I sink myself into the nothingness of God, and experience in eternity, in naught, the 'I',"—just so, the lonely disciple of the late mysteries said to himself: "I listen to the silent universe and fetch[21] the Logos-bearing soul out of the silence. I love the Logos because the Logos brings tidings of an unknown god."

This was an ancient parallel to the admission of Meister Eckhart. Just as the latter immersed himself into the naught of the divine that Medieval theology had proclaimed to him and, out of this void, brought out the 'I,' so that ancient sage listened to a dumb and silent world; for he could no longer hear what traditional wisdom taught him. The spirit-saturated soul had once drawn the ancient wisdom from the universe. This had now turned silent, but still he had a Logos-bearing soul. And he loved the Logos even though it was no longer the godhead of former ages, but only an image of the divine. In other words, already then, the spirit had vanished from the soul's sight. Just as Meister Eckhart later had to seek the 'I' in nothingness, so at that time the soul had to be sought in the despirited world.

Indeed, in former times the souls had the inner firmness needed to say to themselves: In the inward perception of the spirit indwelling me, I myself am something divine. But now, for direct perception, the spirit no longer inhabited the soul. No longer did the soul experience itself as the spirit's messenger, for one must know something in order to

20

be its messenger. Now, the soul only felt itself as the bearer of the Logos, the spirit image; though this spirit image was vivid in the soul. It expressed itself in the love for this god who thus still lived in his image in the soul. But the soul no longer felt like the messenger, only the carrier, of an image of the divine spirit. One can say that a different form of knowledge arose when man looked into his inner being. The soul declined from messenger to bearer.

Soul: Bearer
Body: Force

Since the living spirit had been lost to human perception, the body no longer appeared as the image of spirit. To recognize it as such an image, one would have had to perceive the archetype. Therefore, for this later age, the body changed into something that I would like to call "force." The concept of force emerged. The body was pictured as a complex of forces, no longer as a reproduction, an image, that bore within itself the essence of what it reproduced. The human body became a force which no longer bore the substance of the source from which it originated.

Not only the human body, but in all of nature, too, forces had to be pictured everywhere. Whereas formerly, nature in all its aspects had been an image of spirit, now it had become forces flowing out of the spirit. This, however, implied that nature began to be something more or less foreign to man. One could say that the soul had lost something since it no longer contained direct spirit awareness. Speaking crudely, I would have to say that the soul had inwardly become more tenuous, while the body, the external corporeal world, had gained in robustness. Earlier, as an image, it still possessed some resemblance to the spirit. Now it became permeated by the element of force. The complex of forces is more robust than the image in which the spiritual element is still recognizable. Hence, again speaking crudely, the corporeal world became denser while the soul became more tenuous. This is what arose in the consciousness of the men among whom lived the ancient wise man mentioned above, who listened to the silent universe and, from its silence, derived the awareness that at least his soul was a Logos-bearer.

Now, a contrast that had not existed before arose between the soul, grown more tenuous, and the increased density of the corporeal world. Earlier, the unity of spirit had been perceived in all things. Now, there arose the contrast between body and soul, man and nature. Now appeared a chasm between body and soul that had not been present at all prior to the time of this old sage. Man now felt himself divided as well from nature, something that also had not been the case in the ancient times. This contrast is the central trait of all thinking in the span of time between the old sage I have mentioned and Nicholas Cusanus.

Men now struggle to comprehend the connection between, on one hand, the soul, that lacks spirit reality, and, on the other hand, the body that has become dense, has turned into force, into a complex of forces.

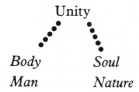

And men struggle to feel and experience the relationship between man and nature. But everywhere, nature is force. In that time, no conception at all existed as yet of what we call today "the laws of nature." People did not think in terms of natural laws; everywhere and in everything they felt the forces of nature. When a man looked into his own being, he did not experience a soul that—as was the case later on—bore within itself a dim will, an almost equally dim feeling, and an abstract thinking. Instead, he experienced the soul as the bearer of the living Logos, something that was not abstract and dead, but a divine living image of God.

We must be able to picture this contrast, which remained acute until the Eleventh or Twelfth century. It was quite different from the contrasts that we feel today. If we cannot vividly grasp this contrast, which was experienced by everyone in that earlier epoch, we make the same mistake as all those historians of philosophy who regard the old Greek thinker Democritus[22] of the fifth century B.C. as an atomist in the modern sense, because he spoke of "atoms." The words suggest a resemblance, but no real resemblance exists. There is great difference between modern-day atomists and Democritus. His utterances were based on the awareness of the contrast described above between man and nature, soul and body. His atoms were complexes of force and as such were contrasted with space, something a modern atomist cannot do in that manner. How could the modern atomist say what Democritus said: "Existence is not more than nothingness, fullness is not more than emptiness?" It implies that Democritus assumed empty space to possess an affinity with atom-filled space. This has meaning only within a consciousness that as yet has no idea of the modern concept of body. Therefore, it cannot speak of the atoms of a body, but only of centers of force, which, in that case, have an inner relationship to what surrounds man externally. Today's atomist cannot equate emptiness with fullness. If Democritus had viewed emptiness the way we do today, he could not have equated it with the state of being. He could do so because in this emptiness he sought the soul that was the bearer of the Logos. And though he conceived this Logos in a form of necessity, it was the Greek form of necessity, not our modern physical necessity. If we are to comprehend what goes on today, we must be able to look in the right way into the nuances of ideas and feelings of former times.

There came the time, described in the last lecture, of Meister Eckhart and Nicholas Cusanus, when even awareness of the Logos indwelling the soul was lost. The ancient sage, in listening to the universe, only had to mourn the silence, but Meister Eckhart and Cusanus found the naught and had to seek the I out of nothingness. Only now, at this point, does the modern era of thinking begin. The soul now no longer contains the living Logos. Instead, when it looks into itself, it finds ideas and concepts, which finally lead to abstractions. The soul has become even more tenuous. A third phase begins. Once upon a time, in the first phase, the soul experienced the spirit's archetype within itself. It saw itself as the messenger of spirit. In the second phase, the soul inwardly experienced the living image of God in the Logos, it became the bearer of the Logos.

Now, in the third phase, the soul becomes, as it were, a vessel for ideas and concepts. These may have the certainty of mathematics, but they are only ideas and concepts. The soul experiences itself at its most tenuous, if I may put it so. Again the corporeal world increases in robustness. This is the third way in which man experiences himself. He cannot as yet give up his soul element completely, but he experiences it as the vessel for the realm of ideas. He experiences his body, on the other hand, not only as a force but as a spatial body.

Soul:	Realm of Ideas
Body:	Spatial Corporeality

The body has become still more robust. Man now denies the spirit altogether. Here we come to the "body" that Hobbes, Bacon,[23] and Locke spoke of. Here, we meet "body" at its

24

densest. The soul no longer feels a kinship to it, only an abstract connection that gets worse in the course of time.

In place of the earlier concrete contrast of soul and body, man and nature, another contrast arises that leads further and further into abstraction. The soul that formerly appeared to itself as something concrete—because it experienced in itself the Logos-image of the divine—gradually transforms itself to a mere vessel of ideas. Whereas before, in the ancient spiritual age, it had felt akin to everything, it now sees itself as subject and regards everything else as object, feeling no further kinship with anything.

The earlier contrast of soul and body, man and nature, increasingly became the merely theoretical epistemological contrast between the subject that is within a person and the object without. Nature changed into the object of knowledge. It is not surprising that out of its own needs knowledge henceforth strove for the "purely objective."

But what is this purely objective? It is no longer what nature was to the Greeks. The objective is external corporeality in which no spirit is any longer perceived. It is nature devoid of spirit, to be comprehended from without by the subject.

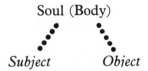

Soul (Body)

Subject   Object

Precisely because man had lost the connection with nature, he now sought a science of nature from outside. Here, we have once again reached the point where I concluded yesterday. Cusanus looked upon what should have been the divine world to him and declared that man with his knowledge must stop short before it and, if he must write about the divine world, he must write a *docta ignorantia*. And only faintly, in symbols taken from mathematics, did Cusanus want to retain something of what appeared thus to him as the spiritual realms.

About a hundred years after the *Docta Ignorantia* appeared in 1440, the *De Revolutionibus Orbium Coelestium* appeared in 1543. One century later, Copernicus, with his mathematical mind, took hold of the other side, the external side of what Cusanus could not fully grasp, not even symbolically, with mathematics. Today, we see how in fact the application of this mathematical mind to nature becomes possible the moment that nature vanishes from man's immediate experience. This can be traced even in the history of language since "Nature" refers to something that is related to "being born," whereas what we consider as nature today is only the corporeal world in which everything is dead. I mean that it is dead for us since, of course, nature contains life and spirit. But it has become lifeless for us and the most certain of conceptual systems, namely, the mathematical, is regarded as the best way to grasp it.

Thus we have before us a development that proceeds with inward regularity. In the first epoch, man beheld god and world, but god in the world and the world in god: the one-ness, unity. In the second epoch, man in fact beheld soul and body, man and nature; the soul as bearer of the living Logos, the bearer of what is not born and does not die; nature as what is born and dies. In the third phase man has ascended to the abstract contrast of subject (himself) and object (the external world). The object is something so robust that man no longer even attempts to throw light on it with concepts. It is experienced as something alien to man, something that is examined from outside with mathematics although mathematics cannot penetrate into the inner essence. For this reason, Cusanus applied mathematics only symbolically, and timidly at that.

The striving to develop science must therefore be pictured as emerging from earlier faculties of mankind. A time had to come when this science would appear. It had to develop the way it did. We can follow this if we focus clearly on the three phases of development that I have just described.

We see how the first phase extends to the Eighth Century B.C., to the ancient sage of Southern Europe whom I have described today. The second extends from him to Nicholas Cusanus. We find ourselves in the third phase now. The first is pneumatological, directed to the spirit in its primeval form. The second is mystical, taking the word in the broadest sense possible. The third is mathematical. Considering the significant characteristics, therefore, we trace the first phase—ancient pneumatology—as far as the ancient Southern wise man. Magical mysticism extends from there to Meister Eckhart and Nicholas Cusanus. The age of mathematizing natural science proceeds from Cusanus into our own time and continues further. More on this tomorrow.

LECTURE III

Dornach, December 26, 1922

In the last two lectures I tried to indicate the point in time when the scientific outlook and manner of thinking, such as we know it today, arose in the course of time. It was pointed out yesterday that the whole character of this scientific thinking, emerging at the beginning most clearly in Copernicus' conception of astronomy, depends on the way in which mathematical thinking was gradually related to the reality of the external world. The development of science in modern times has been greatly affected by a change—one might almost say a revolutionary change—in human perception in regard to mathematical thinking itself. We are much inclined nowadays to ascribe permanent and absolute validity to our own manner of thinking.

Nobody notices how much matters have changed. We take a certain position today in regard to mathematics and to the relationship of mathematics to reality. We assume that this is the way it has to be and that this is the correct relationship. There are debates about it from time to time, but within certain limits this is regarded as the true relationship. We forget that in a none too distant past mankind felt differently concerning mathematics. We need only recall what happened soon after the point in time that I characterized as the most important in modern spiritual life, the point when Nicholas Cusanus presented his dissertation to the world. Shortly after this, not only did Copernicus try to explain the movements of the solar system with mathematically oriented thinking of the kind to which we are accustomed today, but

28

philosophers such as Descartes and Spinoza[24] began to apply this mathematical thought to the whole physical and spiritual universe.

Even in such a book as his *Ethics*, the philosopher Spinoza placed great value on presenting his philosophical principles and postulates, if not in mathematical formulae—for actual calculations play no special part—yet in such a manner that the whole form of drawing conclusions, of deducing the later rules from earlier ones, is based on the mathematical pattern. By and by it appeared self-evident to the men of that time that in mathematics they had the right model for the attainment of inward certainty. Hence they felt that if they could express the world in thoughts arranged in the same clear-cut architectural order as in a mathematical or geometrical system, they would thereby achieve something that would have to correspond to reality. If the character of scientific thinking is to be correctly understood, it must be through the special way in which man relates to mathematics and mathematics relates to reality. Mathematics had gradually become what I would term a self-sufficient inward capacity for thinking. What do I mean by that?

The mathematics existing in the age of Descartes[25] and Copernicus can certainly be described more or less in the same terms as apply today. Take a modern mathematician, for example, who teaches geometry, and who uses his analytical formulas and geometrical concepts in order to comprehend some physical process. As a geometrician, this mathematician starts from the concepts of Euclidean geometry, the three-dimensional space (or merely dimensional space, if he thinks of non-Euclidean geometry).[26] In three-dimensional space he distinguishes three mutually perpendicular directions that are otherwise identical. Space, I would say, is a self-sufficient form that is simply placed before one's consciousness in the manner described above without questions being raised such as: Where does this form come from? or: Where do we get our whole geometrical system?

In view of the increasing superficiality of psychological thinking, it was only natural that man could no longer penetrate to those inner depths of soul where geometrical thought has its base. Man takes his ordinary consciousness for granted and fills this consciousness with mathematics that has been thought-out but not experienced. As an example of what is thought-out but not experienced, let us consider the three perpendicular dimensions of Euclidean space. Man would have never thought of these if he had not experienced a threefold orientation within himself. One orientation that man experiences in himself is from front to back. We need only recall how, from the external modern anatomical and physiological point of view, the intake and excretion of food, as well as other processes in the human organism, take place from front to back. The orientation of these specific processes differs from the one that prevails when, for example, I do something with my right arm and make a corresponding move with my left arm. Here, the processes are oriented to left and right. Finally, in regard to the last orientation, man grows into it during earthly life. In the beginning he crawls on all fours and only gradually stands upright, so that this last orientation flows within him from above downward and up from below.

As matters stand today, these three orientations in man are regarded very superficially. These processes—front to back, right to left or left to right, and above to below—are not inwardly experienced so much as viewed from outside. If it were possible to go back into earlier ages with true psychological insight, one would perceive that these three orientations were inward experiences for the men of that time. Today our thoughts and feelings are still halfway acknowledged as inward experiences, but the man of a by-gone age had a real inner experience, for example, of the front-to-back orientation. He had not yet lost awareness of the decrease in intensity of taste sensations from front to back in the oral cavity. The qualitative experience that taste was strong on the tip of the tongue, then grew fainter and fainter as it receded from front to back, until it disappeared entirely, was once a real and concrete experience. The orientation from front to back was felt in such qualitative experiences. Our inner life is no longer as intense as it once was. Therefore, today, we no longer have experiences such as this. Likewise man today no longer has a vivid feeling for the alignment of his axis of vision in order to focus on a given point by shifting the right axis over the left. Nor does he have a full concrete awareness of what happens when, in the orientation of right-left, he relates his right arm and hand to the left arm and hand. Even less does he have a feeling that would enable him to say: The thought illuminates my head and, moving in the direction from above to below, it strikes into my heart. Such a feeling, such an experience, has been lost to man along with the loss of all inwardness of world experience. But it did once exist. Man did once experience the three perpendicular orientations of space within himself. And these three spatial orientations—right-left, front-back, and above-below—are the basis of the three-dimensional framework of space, which is only the abstraction of the immediate inner experience described above.

So what can we say when we look back at the geometry of earlier times? We can put it like this: It was obvious to a man in those ages that merely because of his being human the geometrical elements revealed themselves in his own life. By extending his own above-below, right-left, and front-back orientations, he grasped the world out of his own being.

Try to sense the tremendous difference between this mathematical feeling bound to human experience, and the bare, bleak mathematical space layout of analytical geometry, which establishes a point somewhere in abstract space, draws three coordinating axes at right angles to each other, and thus isolates this thought-out space scheme from all living experience. But man has in fact torn this thought-out spatial diagram out of his own inner life. So, if we are to understand the origin of the later mathematical way of thinking that was taken over by science, if we are to correctly comprehend its self-sufficient presentation of structures, we must trace it back to the self-experienced mathematics of a bygone age. Mathematics in former times was something completely different. What was once present in a sort of dream-like experience of three-dimensionality and then became abstracted, exists today completely in the unconscious. As a matter of fact, man even now produces mathematics from his own three-dimensionality. But the way in which he derives this outline of space from his experiences of inward orientation is completely unconscious. None of this rises into consciousness except the finished spatial diagram. The same is true of all completed mathematical structures. They have all been severed from their roots. I chose the example of the space scheme, but I could just as well mention any other mathematical category taken from algebra or arithmetic. They are nothing but schemata drawn from immediate human experience and raised into abstraction.

Going back a few centuries, perhaps to the fourteenth century, and observing how people conceived of things mathematical, we find that in regard to numbers they still had an echo of inward feelings. In an age in which numbers had already become as abstract as they are today, people would have been unable to find the names for numbers. The words designating numbers are often wonderfully characteristic. Just think of the word "two" (*zwei*). It clearly expresses a real process, as when we say *entzweien*, "to cleave in twain." Even more, it is related to *zweifeln*, "to doubt." It is not mere imitation of an external process when the number two, *zwei*, is described by the word *Entzweien*, which indicates the disuniting, the splitting, of something formerly a whole. It is in fact something that is inwardly experienced and only then made into a scheme. It is brought up from within, just as the abstract three-dimensional space-scheme is drawn up from inside the mind.

We arrive back at an age of rich spiritual vitality that still existed in the first centuries of Christianity, as can be demonstrated by the fact that mathematics, mathesis, and mysticism were considered to be almost one and the same. Mysticism, mathesis, and mathematics are one, though only in a certain connection. For a mystic of the first Christian centuries, mysticism was something that one experienced more inwardly in the soul. Mathematics was the mysticism that one experienced more outwardly with the body; for example, geometry with the body's orientations to front-and-back, right-and-left, and up-and-down. One could say that actual mysticism was soul mysticism, and that mathematics, mathesis, was mysticism of the corporeality. Hence, proper mysticism was inwardly experienced in what is generally understood by this term; whereas mathesis, the other mysticism, was experienced by means of an inner experience of the body, as yet not lost.

As a matter of fact, in regard to mathematics and the mathematical method Descartes and Spinoza still had completely different feelings from what we have today. Immerse yourself in these thinkers, not superficially as is the practice today when one always wants to discover in the thinkers of old the modern concepts that have been drilled into our heads, but unselfishly, putting yourself mentally in their place. You will find that even Spinoza still retained something of a mystical attitude toward the mathematical method.

The philosophy of Spinoza differs from mysticism only in one respect. A mystic like Meister Eckhart or Johannes Tauler[27] attempts to experience the cosmic secrets more in

33

the depths of feeling. Equally inwardly, Spinoza constructs the mysteries of the universe along mathematical, methodical lines, not specifically geometrical lines, but lines experienced mentally by mathematical methods. In regard to soul configuration and mood, there is no basic difference between the experience of Meister Eckhart's mystical method and Spinoza's mathematical one. Anyone who makes such a distinction does not really understand how Spinoza experienced his *Ethics*, for example, in a truly mathematical-mystical way. His philosophy still reflects the time when mathematics, mathesis, and mysticism were felt as one and the same experience in the soul.

Now, you will perhaps recall how, in my book *The Case for Anthroposophy*,[28] I tried to explain the human organization in a way corresponding to modern thinking. I divided the human organization—meaning the physical one—into the nerve-sense system, the rhythmic system, and the metabolic-limb system. I need not point out to you that I did not divide man into separate members placed side by side in space, although certain academic persons have accused[29] me of such a caricature. I made it clear that these three systems interpenetrate each other. The nerve-sense system is called the "head system" because it is centered mainly in the head, but it spreads out into the whole body. The breathing and blood rhythms of the chest system naturally extend into the head organization, and so on. The division is functional, not local. An inward grasp of this threefold membering will give you true insight into the human being.

Let us now focus on this division for a certain purpose. To begin with, let us look at the third member of the human organization, that of digestion (metabolism) and the limbs. Concentrating on the most striking aspect of this member, we see that man accomplishes the activities of external life by connecting his limbs with his inner experiences. I have characterized some of these, particularly the inward orientation experience of the three directions of space. In his external movements, in finding his orientation in the world, man's limb system achieves inward orientation in the three directions. In walking, we place ourselves in a certain manner into the experience of above-below. In much that we do with our hands or arms, we bring ourselves into the orientation of right-and-left. To the extent that speech is a movement of the aeriform in man, we even fit ourselves into the direction of front-and-back, back-and-front, when we speak. Hence, in moving about in the world, we place our inward orientation into the outer world.

Let us look at the true process, rather than the merely illusionary one, in a specific mathematical case. It is an illusionary process, taking place purely in abstract schemes of thought, when I find somewhere in the universe a process in space, and I approach it as an analytical mathematician in such a way that I draw or imagine the three coordinate axes of the usual spatial system and arrange this external process into Descartes' purely artificial space scheme.

This is what occurs above, in the realm of thought schemes, through the nerve-sense system. One would not achieve a relationship to such a process in space if it were not for what

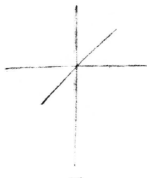

one does with one's limbs, with one's whole body, if it were not for inserting oneself into the whole world in accordance with the inward orientation of above-below, right-left, and front-back. When I walk forward, I know that on one hand I place myself in the vertical direction in order to remain upright. I am also aware that in walking I adjust my direction to the back-to-front orientation, and when I swim and use my arms, I orient myself in right and left. I do not understand all this if I apply Descartes' space scheme, the abstract scheme of the coordinate axes. What gives me the impression of reality in dealing with matters of space is found only when I say to myself: Up in the head, in the nerve system, an illusory image arises of something that occurs deep down in the subconscious. Here, where man cannot reach with his ordinary consciousness, something takes place between his limb system and the universe. The whole of mathematics, of geometry, is brought up out of our limb system of movement. We would not have geometry if we did not place ourselves into the world according to inward orientation. In truth, we geometrize when we lift what occurs in the subconscious into the illusory of the thought scheme. This is the reason why it appears so abstractly independent to us. But this is something that has only come about in recent times. In the age in which mathesis, mathematics, was still felt to be something close to mysticism, the mathematical relationship to all things was also viewed as something human.

Where is the human factor if I imagine an abstract point somewhere in space crossed by three perpendicular directions and then apply this scheme to a process perceived in actual space? It is completely divorced from man, something quite inhuman. This non-human element, which has appeared in recent times in mathematical thinking, was once human. But when was it human?

The actual date has already been indicated, but the inner

aspect is still to be described. When was it human? It was human when man did not only experience in his movements and his inward orientation in space that he stepped forward from behind and moved in such a way that he was aware of his vertical as well as the horizontal direction, but when he also felt the blood's inward activity in all such moving about, in all such inner geometry. There is always blood activity when I move forward. Think of the blood activity present when, as an infant, I lifted myself up from the horizontal to an upright position! Behind man's movements, behind his experience of the world by virtue of movement (which can also be, and at one time was, an inward experience) there stands the experience of the blood. Every movement, small or large, that I experience as I perform it contains its corresponding blood experience. Today blood is to us the red fluid that seeps out when we prick our skin. We can also convince ourselves intellectually of its existence. But in the age when mathematics, mathesis, was still connected with mysticism, when in a dreamy way the experience of movement was inwardly connected with that of blood, man was inwardly aware of the blood. It was one thing to follow the flow of the blood through the lungs and quite another to follow it through the head. Man followed the flow of the blood in lifting his knee or his foot, and he inwardly felt and experienced himself through and through in his blood. The blood has one tinge when I raise my foot, another when I place it firmly on the ground. When I lounge around and doze lazily, the blood's nuance differs from the one it has when I let thoughts shoot through my head. The whole person can take on a different form when, in addition to the experience of movement, he has that of the blood. Try to picture vividly what I mean. Imagine that you are walking slowly, one step at a time; you begin to walk faster; you start to run, to turn yourself, to dance around. Suppose that you were doing all this, not with to-

37

day's abstract consciousness, but with inward awareness:
You would have a different blood experience at each stage,
with the slow walking, then the increase in speed, the running, the turning, the dancing. A different nuance would be
noted in each case. If you tried to draw this inner experience
of movement, you would perhaps have to sketch it like this
(white line). But for each position in which you found yourself during this experience of movement, you would draw a
corresponding inward blood experience (red, blue, yellow)
(see drawing):

Of the first experience, that of movement, you would say
that you have it in common with external space, because you

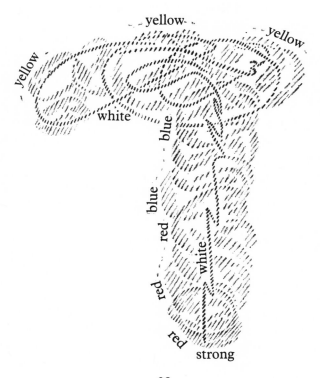

are constantly changing you position. The second experience, which I have marked by means of the different colors, is a time experience, a sequence of inner intense experiences.

In fact, if you run in a triangle, you can have one inner experience of the blood. You will have a different one if you run in a square.

red

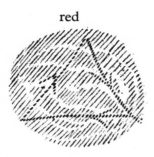

What is outwardly quantitative and geometric, is inwardly intensely qualitative in the experience of the blood.

yellow

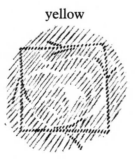

It is surprising, very surprising, to discover that ancient mathematics spoke quite differently about the triangle and square. Modern nebulous mystics describe great mysteries, but there is no great mystery here. It is only what a person would have experienced inwardly in the blood when he walked the outline of a triangle or a square, not to mention the blood experience corresponding to the pentagram. In the blood the whole of geometry becomes qualitative inward experience. We arrive back at a time when one could truly say, as Mephistopheles does in Goethe's *Faust*, "Blood is a very special fluid."[30] This is because, inwardly experienced, the blood absorbs all geometrical forms and makes of them intense inner experiences. Thereby man learns to know himself as well. He learns to know what it means to experience a triangle, a square, a pentagram; he becomes acquainted with the projection of geometry on the blood and its experiences. This was once mysticism. Not only was mathematics, mathesis, closely related to mysticism, it was in fact the external side of movement, of the limbs, while the inward side was the blood experience. For the mystic of bygone times all of mathematics transformed itself out of a sum of spatial formations into what is experienced in the blood, into an intensely mystical rhythmic inner experience.

We can say that once upon a time man possessed a knowledge that he experienced, that he was an integral part of; and that at the point in time that I have mentioned, he lost this oneness of self with the world, this participation in the cosmic mysteries. He tore mathematics loose from his inner being. No longer did he have the experience of movement; instead, he mathematically constructed the relationships of movement outside. He no longer had the blood experience; the blood and its rhythm became something quite foreign to him. Imagine what this implies: Man tears mathematics free from his body and it becomes something abstract. He loses his understanding of the blood experience. Mathematics no longer goes inward. Picture this as a soul mood that arose at a specific time. Earlier, the soul had a different mood than later. Formerly, it sought the connection between blood experience and experience of movement; later, it completely separated them. It no longer related the mathematical and geometrical experience to its own movement. It lost the blood experience. Think of this as real history, as something that occurs in the changing moods of evolution. Verily, a man who lived in the earlier age, when mathesis was still mysticism, put his whole soul into the universe. He measured the cosmos against himself. He lived in astronomy.

Modern man inserts his system of coordinates into the universe and keeps himself out of it. Earlier, man sensed a blood experience with each geometrical figure. Modern man feels no blood experience; he loses the relationship to his own heart, where the blood experiences are centered. Is it imaginable that in the seventh or eighth century, when the soul still felt movement as a mathematical experience and blood as a mystical experience, anybody would have founded a Copernican astronomy with a system of coordinates simply inserted into the universe and totally divorced from man? No, this became possible only when a specific soul constitution arose in evolution. And after that something else became possible as well. The inward blood awareness was lost. Now the time had come to discover the movements of the blood externally through physiology and anatomy. Hence you have this change in evolution: On one hand Copernican astronomy, on the other the discovery of the circulation of the blood by Harvey,[31] a contemporary of Bacon and Hobbes. A world view gained by abstract mathematics cannot produce anything like the ancient Ptolemaic theory, which was essentially bound up with man and the living mathematics he experienced within himself. Now, one experiences an abstract system of coordinates starting with an arbitrary zero point. No longer do we have the inward blood experience; instead, we discover the physical circulation of the blood with the heart in the center.

The birth of science thus placed itself into the whole

41

context of evolution in both its conscious and unconscious processes. Only in this way, out of the truly human element, can one understand what actually happened, what had to happen in recent times for science—so self-evident today—to come into being in the first place. Only thus could it even occur to anybody to conduct such investigations as led, for example, to Harvey's discovery of the circulation of the blood. We shall continue with this tomorrow.

LECTURE IV

Dornach, December 27, 1922

In the last lecture, I spoke of a former view of life from which the modern scientific view has evolved. It still combined the qualitative with the form-related or geometrical elements of mathematics, the qualitative with the quantitative. One can therefore look back at a world conception in which the triangle or another geometrical form was an inner experience, no matter whether the form referred to the surface of a given body or to its path of movement. Geometrical and arithmetical forms were intensely qualitative inner experiences. For example, a triangle and a square were each conceived as emerging from a specific inward experience.

This conception could change into a different one only when men lost their awareness that everything quantitative —including mathematics—is originally experienced by man in direct connection with the universe. It changed when the point was reached where the quantitative was severed from what man experiences. We can determine this moment of separation precisely. It occurred when all concepts of space that included man himself were replaced by the schematic view of space that is customary today, according to which, from an arbitrary starting point, the three coordinates are drawn. The kind of mathematics prevalent today, by means of which man wants to dominate the so-called phenomena of nature, arose in this form only after it had been separated from the human element. Expressing it more graphically, I would say that in a former age man perceived mathematics as something that he experienced within himself together with his god or gods, whereby the god ordered the world. It came as no surprise therefore to discover this mathematical order in the world. In contrast to this, to impose an arbitrary space outline or some other mathematical formula on natural phenomena—even if such abstract mathematical concepts can be identified with significant aspects in these so-called natural phenomena—is a procedure that cannot be firmly related to human experiences. Hence, it cannot be really understood and is at most simply assumed to be a fact. Therefore in reality it cannot be an object of any perception. The most that can be said of such an imposition of mathematics on natural phenomena is that what has first been mathematically thought out is then found to fit the phenomena of nature. But why this is so can no longer be discovered within this particular world perception.

Think back to the other worldview that I have previously described to you, when all corporeality was regarded as image of the spirit. One looking at a body found in it the image of spirit. One then looked back on oneself, on what—in union with one's own divine nature—one experienced as mathematics through one's own bodily constitution. As a work of art is not something obscure but is recognized as the image of the artist's ideas, so one found in corporeal nature the mathematical images of what one had experienced with one's own divine nature. The bodies of external nature were images of the divine spiritual. The instant that mathematics is separated from man and is regarded only as an attribute of bodies that are no longer seen as a reflection of spirit, in that instant agnosticism creeps into knowledge.

Take a concrete example, the first phenomenon that confronts us after the birth of scientific thinking, the Copernican system. It is not my intention today or in any of these lectures to defend either the Ptolemaic or the Copernican system. I am not advocating either one. I am only speaking of the historical fact that the Copernican system has replaced

44

the Ptolemaic. What I say today does not imply that I favor the old Ptolemaic system over the Copernican. But this must be said as a matter of history. Imagine yourself back in the age when man experienced his own orientation in space: above-below, right-left, front-back. He could experience this only in connection with the earth. He could, for example, experience the vertical orientation in himself only in relation to the direction of gravity. He experienced the other two in connection with the four compass points according to which the earth itself is oriented. All this he experienced *together* with the earth as he felt himself standing firmly on it. He thought of himself not just as a being that begins with the head and ends at the sole of the feet. Rather, he felt himself penetrated by the force of gravity, which had something to do with his being but did not cease at the soles of his feet. Hence, feeling himself within the nature of the gravitational force, man felt himself one with the earth. For his concrete experience, the starting point of his cosmology was thus given by the earth. Therefore he felt the Ptolemaic system to be justified.

Only when man severed himself from mathematics, only then was it possible also to sever mathematics from the earth and to found an astronomical system with its center in the sun. Man had to lose the old experience-within-himself before he could accept a system with its center outside the earth. The rise of the Copernican system is therefore intimately bound up with the transformation of civilized mankind's soul mood. The origin of modern scientific thinking cannot be separated from the general mental and soul condition, but must be viewed in context with it.

It is only natural that statements like this are considered absurd by our contemporaries, who believe in the present world view far more fervently than the sectarians of olden days believed in their dogmas. But to give the scientific mode of thinking its proper value, it must be seen as arising

45

inevitably out of human nature and evolution. In the course of these lectures, we shall see that by doing this we are actually assigning far greater value to science than do the modern agnostics.

Thus the Copernican world conception came into being, the projection of the cosmic center from the earth to the sun. Fundamentally, the whole cosmic thought edifice of Giordano Bruno,[32] who was born in 1548 and burned at the stake in Rome in 1600, was already contained in the Copernican world view. It is often said that Giordano Bruno glorifies the modern view of nature, glorifies Copernicanism. One must have deep insight into the inner necessity with which this new cosmology arose if one is to have any feeling at all for the manner and tone in which Giordano Bruno speaks and writes. Then one sees that Giordano Bruno does not sound like the followers of the new view or like the stragglers of the old view. He really does not speak about the cosmos mathematically so much as lyrically. There is something musical in the way Giordano Bruno describes the modern conception of nature. Why is that? The reason is that Giordano Bruno, though he was rooted with his whole soul in a bygone world perception, told himself with his outward intellect: The way things have turned out in history, we cannot but accept the Copernican world picture. He understood the absolute necessity that had been brought about by evolution. This Copernican world view, however, was not something he had worked out for himself. It was something given to him, and which he found appropriate for his contemporaries. Belonging as he did to an older world conception, he could not help but experience inwardly what he had to perceive and accept as knowledge. He still had the faculty of inner experience, but he did not have scientific forms for it. Therefore although he described them so wonderfully, he did not follow the Copernican directions of thought in the manner of Copernicus, Galileo, Kepler, or

Newton.[33] Instead, he tried to experience the cosmos in the old way, the way that was suitable when the whole cosmos was experienced within one's being. But in order to do this, mathematics would have had to be also mysticism, inward experience, in the way I described yesterday. This it could not be for Giordano Bruno. The time for it was past. Hence, his attempt to enter the new cosmology through living experience became an experience, not of knowledge but of poetry, or at least partially so. This fact lends Giordano's works their special coloring. The atom is still a monad; in his writings, it is still something alive. The sum of cosmic laws retains a soul quality, but not because he experienced the soul in all the smallest details as did the ancient mystics, and not because he experienced the mathematical laws of the cosmos as the intentions of the spirit. No, it was because he roused himself to wonder at this new cosmology and to glorify it poetically in a pseudo-scientific form. Giordano Bruno is truly something like a connecting link between two world conceptions, the present one and the ancient one that lasted into the fifteenth century. Man today can form scarcely any idea of the latter. All cosmic aspects were then still experienced by man, who did not yet differentiate between the subject within himself and the cosmic object outside. The two were still as one; man did not speak of the three dimensions in space, sundered from the orientation within his own body and appearing as above-below, right-left, and forward-backward.

Copernicus tried to grasp astronomy with abstract mathematical ideas. On the other hand, Newton shows mathematics completely on its own. Here I do not mean single mathematical deductions, but mathematical thinking in general, entirely divorced from human experience. This sounds somewhat radical and objections could certainly be made to what I am thus describing in broad outlines, but this does not alter the essential facts. Newton is pretty much the first to approach the phenomena of nature with abstract mathematical thinking. Hence, as a kind of successor to Copernicus, Newton becomes the real founder of modern scientific thinking.

It is interesting to see in Newton's time and in the age that followed how civilized humanity is at pains to come to terms with the immense transformation in soul configuration that occurred as the old mathematical-mystical view gave way to the new mathematical-scientific style. The thinkers of the time find it difficult to come to terms with this revolutionary change. It becomes all the more evident when we look into the details, the specific problems with which some of these people wrestled. See how Newton, for instance, presents his system by trying to relate it to the mathematics that has been severed from man. We find that he postulates time, place, space, and motion. He says in effect in his *Principia*: I need not define place, time, space, and motion because everybody understands them.[34] Everybody knows what time is, what space, place, and motion are, hence these concepts, taken from common experience, can be used in my mathematical explanation of the universe. People are not always fully conscious of what they say. In life, it actually happens seldom that a person fully penetrates everything he says with his consciousness. This is true even among the greatest thinkers. Thus Newton really does not know why he takes place, time, space, and motion as his starting points and feels no need to explain or define them, whereas in all subsequent deductions he is at pains to explain and define everything. Why does he do this? The reason is that in regard to place, time, motion, and space all cleverness and thinking avail us nothing. No matter how much we think about these concepts, we grow no wiser than we were to begin with. Their nature is such that we experience them simply through our common human nature and must take them as they come. A successor of Newton's,

Bishop Berkeley,[35] took particular notice of this point. He was involved in philosophy more than Newton was, but Berkeley illustrates the conflicts taking place during the emergence of scientific thinking. In other respects, as we shall presently hear, he was not satisfied with Newton, but he was especially struck by the way that Newton took these concepts as his basis without any explanation, that he merely said: I start out from place, time, space, and motion; I do not define them; I take them as premises for my mathematical and scientific reflections. Berkeley agrees that one must do this. One must take these concepts in the way they are understood by the simplest person, because there they are always clear. They become unclear not in outward experience, but in the heads of metaphysicians and philosophers. Berkeley feels that when these four concepts are found in life, they are clear; but they are always obscure when found in the heads of thinkers.

It is indeed true that all thinking about these concepts is of no avail. One feels this. Therefore, Newton is only beginning to juggle mathematically when he uses these concepts to explain the world. He is juggling with ideas. This is not meant in a derogatory way; I only want to describe Newton's abilities in a telling manner. One of the concepts thus utilized by Newton is that of space. He manipulates the idea of space as perceived by the man in the street. Still, a vestige of living experience is contained therein. If, on the other hand, one pictures space in terms of Cartesian mathematics without harboring any illusions, it makes one's brain reel. There is something undefinable about this space, with its arbitrary center of coordinates. One can, for example, speculate brilliantly (and fruitlessly) about whether Descartes' space is finite or infinite. Ordinary awareness of space that is still connected with the human element really is not at all concerned with finiteness or infinity. It is after all quite without interest to a living world conception whether space can be pictured as finite or infinite. Therefore one can say that Newton takes the trivial idea of space just as he finds it, but then he begins to mathematize. But, due to the particular quality of thinking in his age, he already has the abstracted mathematics and geometry, and therefore he penetrates spatial phenomena and processes of nature with abstract mathematics. Thereby he sunders the natural phenomena from man. In fact, in Newton's physics we meet for the first time ideas of nature that have been completely divorced from man. Nowhere in earlier times were conceptions of nature so torn away from man as they are in Newtonian physics.

Going back to a thinker of the fourth or fifth century A.D.—though people of that period can hardly be called "thinkers" because their inner life was far more alive than the mere life in thoughts—we would find that he held the view: "I live; I experience space along with my God. I orient myself in space up-and-down, right-left, and forward-backward, but I dwell in space together with my God. He outlines the directions and I experience them." So it was for a thinker of the third or fourth century A.D. and even later; indeed, it only became different in the fourteenth century. Thinking geometrically about space, man did not merely draw a triangle but was conscious of the fact that, while he did this, God dwelled within him and drew along with him. His experience was qualitative; he drew the qualitative reality that God Himself had placed within him. Everywhere in the outer world, whenever mathematics was observed, the intentions of God were also observed.

By Newton's time mathematics has become abstracted. Man has forgotten that originally he received mathematics as an inspiration from God. And in this utterly abstract form, Newton now applies mathematics to the study of space. As he writes his *Principia*, he simply applies this abstracted mathematics, this idea of space (which he does not define), becaue he has a dim feeling that nothing will be gained by trying to define it. He takes the trivial idea of space and applies his abstract mathematics to it, thus severing it from any inward experiences. This is how he speaks of the principles of nature.

Later on, interestingly enough, Newton goes somewhat deeper. This is easy to see if one is familiar with his works. Newton becomes ill at ease, as it were, when he contemplates his own view of space. He is not quite comfortable with this space, torn as it is out of man and estranged completely from the spirit. So he defines it after all, saying that space is the sensorium of God. It is most interesting that at the starting point of modern science the very person who was the first to completely mathematize and separate space from man, eventually defines space as God's sensorium,[36] a sort of brain or sense organ of God. Newton had torn nature asunder into space and man-who-experiences-space. Having done this, he feels inwardly uneasy when he views this abstract space, which man had formerly experienced in union with his god. Formerly, man had said to himself: What my human sensorium experiences in space, I experience together with my god. Newton becomes uneasy, now that he has torn space away from the human sensorium. He has thereby torn himself away from his permeation with the divine-spiritual. Space, with all its mathematics, was now something external. So, in later life, Newton addresses it as God's sensorium, though to begin with he had torn the whole apart, thus leaving space devoid of Spirit and God. But enough feeling remained in Newton that he could not leave this externalized space devoid of God. So he deified it again.

Scientifically, man tore himself loose from his god, and thus from the spirit; but outwardly he again postulated the same spirit. What happened here explains why a man like Goethe found it impossible[37] to go along with Newton on any point. Goethe's *Theory of Color* is one particularly characteristic point. This whole procedure of first casting out the spirit, separating it from man, was foreign to Goethe's nature. Goethe always had the feeling that man has to experience everything, even what is related to the cosmos. Even in regard to the three dimensions Goethe felt that the cosmos was only a continuation of what man had inwardly experienced. Therefore Goethe was by nature Newton's adversary.

Now let us return to Berkeley, who was somewhat younger than Newton, but still belonged to the period of conflict that accompanied the rise of the scientific way of thinking. Berkeley had no quarrel with Newton's accepting the trivial ideas of place, space, time, and motion. But he was not happy with this whole science that was emerging, and particularly not with its interpretations of natural phenomena. It was evident to him that when nature is utterly severed from man it cannot be experienced at all, and that man is deceiving himself when he imagines that he is experiencing it.

Therefore, Berkeley declared that bodies forming the external basis for sense perceptions do not really exist. Reality is spiritual through and through. The universe, as it appears to us—even where it appears in a bodily form—is but the manifestation of an all-pervading spirit. In Berkeley, these ideas appear pretty much as mere assertions, for he no longer had any trace of the old mysticism and even less of the ancient pneumatology. Except for his religious dogma, he really had no ground at all for his assertion of such all-pervading spirituality. But assert it he did, and so vigorously that all corporeality became for him no more than a revelation of the spirit. Hence it was impossible for Berkeley to say: I behold a color and there is vibrating movement back of it that I cannot see—which is what modern science justifiably states. Instead, Berkeley said: I cannot hypothetically assume that there is anything possessing any corporeal property such as vibratory movement. The basis of the physical world of phenomena must be spiritually conceived. Something spiritual is behind a color perception as its cause, which I experience in myself when I know myself as spirit. Thus Berkeley is a spiritualist in the sense in which this term is used in German philosophy.

For dogmatic reasons, but with a certain justification, Berkeley makes innumerable objections against the assumption that nature can be comprehended by a mathematics that has been abstracted from direct experience. Since to Berkeley the whole cosmos was spiritual, he also viewed mathematics as having been formed together with the spirit of the cosmos. He held that we do in fact experience the intentions of the cosmic spirit insofar as they have mathematical form, but that we cannot apply mathematical concepts in an external manner to corporeal objects.

In accordance with this point of view, Berkeley opposed what mathematics had become for both Newton and Leibnitz,[38] namely differential and integral calculus. Please, do not misunderstand me. Today's lecture must be fashioned in such a way that it cannot but provoke many objections in one who holds to the views prevailing today. But these objections will fade away during the ensuing lectures, if one is willing to keep an open mind. Today, however, I want to present the themes that will occupy us in a rather radical form.

Berkeley became an opponent of the whole infinitesimal calculus[39] to the extent that it was then known. He opposed what was beyond experience. In this regard, Berkeley's feeling for things was often more sensitive than his thoughts. He felt how, to the quantities that the mind could conceive, the emergence of infinitesimal calculus added other quantities; namely, the differentials, which attain definition only in the differential coefficient. Differentials must be conceived

53

in such a way that they always elude our thinking, as it were. Our thinking refuses to completely permeate them. Berkeley regarded this as a loss of reality, since knowledge for him was only what could be experienced. Therefore he could not approve of mathematical ideas that produced the indetermination of the differentials.

What are we really doing when we seek differential equations for natural phenomena? We are pointing to something that eludes our possible experience. I realize, of course, that many of you cannot quite follow me on these points, but I cannot here expound the whole nature of infinitesimal calculus. I only want to draw attention to some aspects that will contribute to our study of the birth of modern science.

Modern science set out to master the natural phenomena by means of a mathematics detached from man, a mathematics no longer inwardly experienced. By adopting this abstract mathematical view and these concepts divorced from man, science arrived at a point where it could examine only the inanimate. Having taken mathematics out of the sphere of live experience, one can only apply it to what is dead. Therefore, owing to this mathematical approach, modern science is directed exclusively to the sphere of death. In the universe, death manifests itself in disintegration, in atomization, in reduction to microscopic parts— putting it simply, in a crumbling into dust. This is the direction taken by the present-day scientific attitude. With a mathematics detached from all living experience, it takes hold of everything in the cosmos that turns to dust, that atomizes. From this moment onward it becomes possible to dissipate mathematics itself into differentials. We actually kill all living forms of thought, if we try to penetrate them with any kind of differential equation, with any differential line of thought. To differentiate is to kill; to integrate is to piece the dead together again in some kind of framework, to fit the differentials together again into a whole. But they do not thereby become alive again, after having been annihilated. One ends up with dead spectres, not with anything living.

This is how the whole perspective of what was opening up through infinitesimal calculus appeared to Berkeley. Had he expressed himself concretely, he might well have said: First you kill the whole world by differentiating it; then you fit its differentials together again in integrals, but you no longer have a world, only a copy, an illusion. With regard to its content, every integral is really an illusion, and Berkeley already felt this to be so. Therefore, differentiation really implies annihilation, while integration is the gathering up of bones and dust, so that the earlier forms of the slain beings can be pieced together again. But this does not bring them back to life; they remain no more than dead replicas.

One can say that Berkeley's sentiments were untimely. This they certainly were, for the new way of approach had to come. Anyone who would have said that infinitesimal calculus should never have been developed would have been called not a scientific thinker but a fool. On the other hand, one must realize that at the outset of this whole stream of development, feelings such as Berkeley's were understandable. He shuddered at what he thought would come from an infinitesimal study of nature; no longer a study of what formerly was considered nature and had to do with the process of birth but a study of all dying aspects in nature.

Formerly this had not been observed, nor had there been any interest in it. In earlier times, the coming-into-being, the germinating, had been studied; now, one looked at all that was fading and crumbling into dust. Man's conception was heading toward atomism, whereas previously it had tended toward the continuous, lasting aspects of things. Since life cannot exist without death and all living things must die, we must look at and understand all that is dead in the world. A science of the inanimate, the dead, had to

55

arise. It was absolutely necessary. The time that we are speaking about was the age in which mankind was ready for such a science. But we must visualize how this went against the grain of somebody who, like Berkeley, still lived completely in the old view.

The after-effects of what came into being then are still very much with us today. We have witnessed the triumphs of just those scientific labors that made Berkeley shudder. Until they were somewhat modified through the modern theory of relativity,[40] Newton's theories reigned supreme, Goethe's revolt against them made no impression. For a true comprehension of what went on we must go back to Newton's time and see the shuddering of thinkers who still had a vivid recollection of earlier views and how they clung to feelings that resembled the former ones.

Giordano Bruno shrank from studying the dead nature that was now to be the object of study. He could not view it as dead in a purely mathematical manner of thought, so he animated the atoms into monads and imbued his mathematical thinking with poetry in order to retain it in a personal sphere. Newton at first proceeded from a purely mathematical standpoint, but then he wavered and defined space (which he has first completely divorced from man through his external mathematics) as God's sensorium. Berkeley in his turn rejected the new direction of thinking altogether and with it the whole trend towards the infinitesimal.

Today, however, we are surrounded and overwhelmed by the world view that Giordano Bruno tried to turn into poetry, that Newton felt uncomfortable about, and that Berkeley completely rejected. Do we take what Newton said—that space is a sensorium of God—seriously when we think in the accepted scientific sense today? People today like to regard as great thinkers those men who have said something or other that they approve. But if the great men also said something that they do not approve, they feel very superior and think: Unfortunately, on this point he wasn't as enlightened as I am. Thus many people consider Lessing[41] a man of great genius but make an exception for what he did towards the end of his life, when he became convinced that we go through repeated earth lives.

Just because we must in the present age come to terms with the ideas that have arisen, we must go back to their origin. Since mathematics has once and for all been detached from man, and since nature has been taken hold of by this abstract mathematics that has gradually isolated us from the whole of nature, we must now somehow manage to find ourselves in this nature. For we will not attain a coherent spiritual knowledge until we once again have found the spirit in nature.

Just as it is a matter of course that every living man will sooner or later die, so it was a matter of course that sooner or later in the course of time a conception of death had to emerge from the former life-imbued world view. Things that can only be learned from a corpse cannot be learned by a person who is unwilling to examine the corpse. Therefore certain mysteries of the world can be comprehended only if the modern scientific way of thinking is taken seriously.

Let me close with a somewhat personal remark.[42] The scientific world view must be taken seriously, and for this reason I was never an opponent of it; on the contrary, I regarded it as something that of necessity belongs to our time. Often I had to speak out against something that a scientist, or so-called scientist, had made of the things that were discovered by unprejudiced investigation of the sphere of death. It was the misinterpretation of such scientific discoveries that I opposed. On this occasion let me state emphatically that I do not wish to be regarded as in any way an opponent of the scientific approach. I would consider it detrimental to all our anthroposophical endeavors if a false opposition were to arise between what anthroposophy seeks by way of spiritual research and what science seeks—and must of necessity seek in its field—out of the modern attitude.

I say this expressly, my dear friends, because a healthy discussion concerning the relationship between anthroposophy and science must come to pass within our movement. Anything that goes wrong in this respect can only do grave harm to anthroposophy and should be avoided.

I mention this here because recently, in preparing these lectures, I read in the anthroposophical periodical *Die Drei* that atomism was being studied in a way in which no progress can be made. Therefore, I want to make it clear that I consider all these polemics in *Die Drei* about atomism as something that only serves to stultify the relations between anthroposophy and science.

LECTURE V

Dornach, December 28, 1922

The isolation of man's ideas (especially his mathematical ideas) from his direct experience has proved to be the outstanding feature of the spiritual development leading to modern scientific thinking. Let us place this process once more before our mind's eye.

We were able to look back into ages past, when what man had to acquire as knowledge of the world was experienced in communion with the world. During those epochs, man inwardly did not experience his threefold orientation— up-down, left-right, front-back—in such a manner that he attributed it solely to himself. Instead, he felt himself within the universal whole; hence, his own orientations were to him synonymous with the three dimensions of space. What he pictured of knowledge to himself, he experienced jointly with the world. Therefore, with no uncertainty in his mind, he knew how to apply his concepts, his ideas, to the world. This uncertainty has only arisen along with the more recent civilization. We see it slowly finding its way into the whole of modern thought and we see science developing under this condition of uncertainty. This state of affairs must be clearly recognized.

A few examples can illustrate what we are dealing with. Take a thinker like John Locke, who lived from the seventeenth into the eighteenth century. His writings show what an up-to-date thinker of his age had to say concerning the scientific world perception. John Locke[43] divided everything that man perceives in his physical environment into two aspects. He divided the characteristic features of bodies into primary and secondary qualities. Primary qualities were those that he could only attribute to the objects themselves, such as shape, position, and motion. Secondary qualities in his view were those that did not actually belong to the external corporeal things but were an effect that these objects had upon man. Examples are color, sound, and warmth. Locke stated it thus: "When I hear a sound, outside of me there is vibrating air. In a drawing, I can picture these vibrations in the air that emanate from a sound-aroused body and continue on into my ear. The shape that the waves, as they are called, possess in the vibrating air can be pictured by means of spatial forms. I can visualize their course in time, hence in movement. Whatever is occurring in space—the shape, movement, and determination of position of things—all this, belonging to the primary qualities, certainly exists in the external world, but it is silent, it is soundless. The quality of sound, a secondary quality, only arises when the vibration of the air strikes my ear, and with it arises that peculiar inner experience that I carry within me as sound. It is the same with color, which is now lumped together with light. There must be something out there in the world that is somehow of a corporeal nature and somehow possesses shape and movement. This exercises an effect on my eye and thus becomes my experience of light or color. It is the same with the other things that present themselves to my senses. The whole corporeal world must be viewed like this; we must distinguish between the primary qualities in it, which are objective, and the secondary qualities, which are subjective and are the effects of the primary qualities upon us."

Simply put, one could say with Locke that the external world outside of man is form, position, and movement, whereas all that makes up the content of the sense world exists in truth somehow inside us. The actual content of color as a human experience is nowhere in the environment, it lives in me. The actual content of sound is nowhere to be found outside, it lives in me. The same is true of my experience of warmth or cold.

In former ages, when what had become the content of knowledge was experienced jointly with the world, one could not possibly have had this view because, as I have said, man experienced mathematics by participating in his own bodily orientation and placing this orientation into his own movement. He experienced this, however, in communion with the world. Therefore, his own experience was sufficient reason for assuming the objectivity of position, place, and movement. Also, though in another portion of his inner life, man again had this communion with the world in regard to color, tone, and so forth. Just as the concept of movement was gained through the experience of his own movement, so the concept of color was gained through a corresponding internal experience in the blood, and this experience was then connected with whatever is warmth, color, sound, and so forth in the surrounding world. Certainly, in earlier times, man distinguished position, location, movement, and time-sequence from color, sound, and warmth, but these were distinguished as being different kinds of experiences that were undergone jointly with different kinds of existence in the objective world.

Now, in the scientific age, the determination of place, movement, position, and form ceased to be inward self-experience. Instead, they were regarded as mere hypotheses that were caused by some external reality. When the shape of a cannon is imagined, one can hardly say: This form of the cannon is actually somehow within me. Therefore its identification was directed outward and the imagined form of the cannon was related to something objective. One could not very well admit that a musket-ball was actually flying within one's brain; therefore, the hypothetically thought-out movements were attributed to something objective.

On the other hand, what one saw in the flying musket-ball, the flash by which one perceived it and the sound by which one heard it, were pushed into one's own human nature, since no other place could be found for them. Man no longer knew how he experienced them jointly with the objects; therefore, he associated them with his own being.

It actually took quite some time before those who thought along the lines of the scientific age perceived the impossibility of this arrangement. What had in fact taken place? The secondary qualities, sound, color, and warmth experience, had become, as it were, fair game in the world and, in regard to human knowledge, had to take refuge in man. But before too long, nobody had any idea of how they lived there. The experience, the self-experience, was no longer there. There was no connection with external nature, because it was not experienced anymore. Therefore these experiences were pushed into one's self. So far as knowledge was concerned, they had, as it were, disappeared inside man. Vaguely it was thought that an ether vibration out in space translated itself into form and movement, and this had an effect on the eye, and then worked on the optic nerve, and finally somehow entered the brain. Our thoughts were a means of looking around inside for whatever it was that, as an effect of the primary qualities, supposedly expressed itself in man as secondary qualities. It took a long time, as I said, before a handful of people firmly pointed out the oddity of these ideas. There is something extraordinary in what the Austrian philosopher Richard Wahle[44] wrote in his *Mechanism of Thinking*, though he himself did not realize the full implications of his sentence: "Nihil est in cerebro, quod non est in nervis" ("There is nothing in the brain that is not in the nerves"). It may not be possible with the means available today to examine the nerves in every conceivable way, but even if we could we would not find sound, color, or warmth experience in them. Therefore, they must not be in the brain either. Actually, one has to admit now that they simply disappear insofar as knowledge is concerned. One examines the relationship of man to the world. Form, position, place, time, etc. are beheld as objective. Sound, warmth experience, and color vanish; they elude one.[45]

Finally, in the Eighteenth Century, this led Kant[46] to say that even the space and time qualities of things cannot somehow be outside and beyond man. But there had to be some relationship between man and the world. After all, such a relationship cannot be denied if we are to have any idea of how man exists together with the world. Yet, the common experience of man's space and time relationships with the world simply did not exist anymore. Hence arose the Kantian idea: If man is to apply mathematics, for example, to the world, then it is *his* doing that he himself makes the world into something mathematical. He impresses the whole mathematical system upon the "things in themselves," which themselves remain utterly unknown.—In the Nineteenth Century science chewed on this problem interminably. The basic nature of man's relation to cognition is simply this: uncertainty has entered into his relationship with the world. He does not know how to recognize in the world what he is experiencing. This uncertainty slowly crept into all of modern thinking. We see it entering bit by bit into the spiritual life of recent times.

It is interesting to place a recent example side by side with Locke's thinking. August Weismann,[47] a biologist of the Nineteenth Century, conceived the thought: in any living organism, the interplay of the organs (in lower organisms, the interaction of the parts) must be regarded as the essential thing. This leads to comprehension of how the organism lives. But in examining the organism itself, in understanding it through the interrelationship of its parts, we find no equivalent for the fact that the organism must die. If one only observes the organism, so Weismann said, one finds nothing that will explain death. In the living organism, there is absolutely nothing that leads to the idea that the organism must die. For Weismann, the only thing that demonstrates that an organism must die is the existence of a corpse. This means that the concept of death is not gained from the living organism. No feature, no characteristic, found in it indicates that dying is a part of the organism. It is only when the event occurs, when we find a corpse in the place of the living organism, that we know the organism possesses the ability to die.

But, says Weismann, there is a class of organisms where corpses are never found. These are the unicellular organisms. They only divide themselves so there are no corpses. The propagation of such beings looks like this:

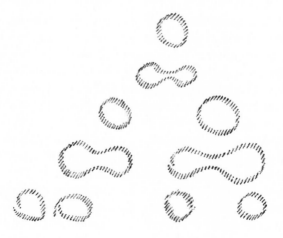

One divides into two; each of these divide into two again, and so on. There is never a corpse. Weismann therefore concludes that the unicellular beings are immortal. This is the immortality of unicellular beings that was famous in nineteenth-century biology. Why were these organisms considered immortal? Because they never produce any corpses, and because we cannot entertain the concept of death in the organic realm as long as there are no corpses. Where there is no corpse, there is no room for the concept of death. Hence, living beings that produce no corpses are immortal.

This example shows how far man has removed himself in modern times from any connection between the world and his thinking, his inner experiences. His concept of an organism is no longer such that the fact of its death can be perceived from it. This can only be deduced from the existence of something like a corpse. Certainly, if a living organism is only viewed from outside, if one cannot experience what is in it, then indeed one cannot find death in the organism and an external sign is necessary. But this only proves that in his thinking man feels himself separated from the things around him.

From the uncertainty that has entered all thinking concerning the corporeal world, from this divorce between our thoughts and our experience, let us turn back to the time when self-experience still existed. Not only did the inwardly experienced concept exist alongside the externally excogitated concept of a triangle, square, or pentagram, but there were also inwardly experienced concepts of blossoming and fading, of birth and death. This inner experience of birth and death had its gradations. When a child was seen to grow more and more animated, when its face began to express its soul, when one really entered into this growing process of the child, this could be seen as a continuation of the process of birth, albeit a less pronounced and intensive one. There were degrees in the experience of birth. When a man began to show wrinkles and grey hair and grew feeble, this was seen as a first mild degree of dying. Death itself was only the sum total of many less pronounced death experiences, if I may use such a paradox. The concepts of blossoming and decaying, of being born and dying, were inwardly alive.

These concepts were experienced in communion with the corporeal world. No line was drawn between man's self-experience and the events in nature. Without a coastline, as it were, the inner land of man merged into the ocean of the universe. Owing to this form of experience, man lived himself into the world itself. Therefore, the thinkers of earlier ages, whose ideas no longer receive proper attention from science, had to form quite different ideas concerning something like what Weismann called the "immortality of unicellular beings." What sort of concept would an ancient thinker have formed had he had a microscope and known something about the division of unicellular organisms? He would have said: First I have the unicellular being; it divides itself into two. Somewhat imprecisely, he might have said: It atomizes itself, it divides itself; for a certain length of time, the two parts are indivisible; then they divide again. As soon as division or atomization begins, death enters in.—He would not have derived death from the corpse but from atomization, from the division into parts. His train of thought would have been somewhat as follows: A being that is capable of life, that is in the process of growth, is not atomized; and when the tendency to atomization appears, the being dies. In the case of unicellular beings, he would simply have thought that the two organisms cast off by the first unicellular being were for the moment dead, but would be, so to speak, revived immediately, and so forth. With atomization, with the process of splitting, he would have linked the thought of death. If he had known about unicellular beings and had seen one split into two, he would not have thought that two new ones had come into being. On the contrary, he would have said that out of the living monad, two atoms have originated. Further, he would have said that wherever there is life, wherever one observes life, one is not dealing with atoms. But if they are found in a living being, then a proportionate part of the being is dead. Where atoms are found, there is death, there is something inorganic. This is how matters would have been judged in a former age based on living inner knowledge of the world.

All this is not clearly described in our histories of philosophy, although the discerning reader can have little doubt of it. The reason is that the thought-forms of this older philosophy are totally unlike today's thinking. Therefore anyone writing history nowadays is apt to put his own modern concepts into the minds of earlier thinkers.[48] But this is impermissible even with a man as recent as Spinoza. In his book on what he justifiably calls ethics, Spinoza follows a mathematical method but it is not mathematics in the modern sense. He expounds his philosophy in a mathematical style, joining idea to idea as a mathematician would. He still retains something of the former qualitative experience of quantitative mathematical concepts. Hence, even in contemplating the qualitative aspect of man's inner life, we can say that his style is mathematical. Today with our current concepts, it would be sheer nonsense to apply a mathematical style to psychology, let alone ethics.

If we want to understand modern thinking, we must continually recall this uncertainty, contrasting it to the certainty that existed in the past but is no longer suited to our modern outlook. In the present phase of scientific thinking, we have come to the point where this uncertainty is not only recognized but theoretical justifications have been offered for it. An example is a lecture given by the French thinker Henri Poincaré[49] in 1912 on current ideas relating to matter. He speaks of the existing controversy or debate concerning the nature of matter; whether it should be thought of as being continuous or discrete; in other words, whether one should conceive of matter as substantial essence that fills

67

space and is nowhere really differentiated in itself, or whether substance, matter, is to be thought of as atomistic, signifying more or less empty space containing within it minute particles that by virtue of their particular interconnections form into atoms, molecules, and so forth.

Aside from what I might call a few decorative embellishments intended to justify scientific uncertainty, Poincaré's lecture comes down to this: Research and science pass through various periods. In one epoch, phenomena appear that cause the thinker to picture matter in a continuous form, making it convenient to conceive of matter this way and to focus on what shows up as continuity in the sense data. In a different period the findings point more toward the concept of matter being diffused into atoms, which are pictured as being fused together again; i.e., matter is not continuous but discrete and atomistic. Poincaré is of the opinion that always, depending on the direction that research findings take, there will be periods when thinking favors either continuity or atomism. He even speaks of an oscillation between the two in the course of scientific development. It will always be like this, he says, because the human mind has a tendency to formulate theories concerning natural phenomena in the most convenient way possible. If continuity prevails for a time, we get tired of it. (These are not Poincaré's exact words, but they are close to what he really intends). Almost unconsciously, as it were, the human mind then comes upon other scientific findings and begins to think atomistically. It is like breathing where exhalation follows inhalation. Thus there is a constant oscillation between continuity and atomism. This merely results from a need of the human mind and according to Poincaré, says nothing about the things themselves. Whether we adopt continuity or atomism determines nothing about things themselves. It is only our attempt to come to terms with the external corporeal world.

It is hardly surprising that uncertainty should result from an age which no longer finds self-experience in harmony with what goes on in the world but regards it only as something occurring inside man. If you no longer experience a living connection with the world, you cannot experience continuity or atomism. You can only force your preconceived notions of continuity or atomism on the natural phenomena. This gradually leads to the suspicion that we formulate our theories according to our changing needs. Just as we must breath in and out, so we must, supposedly, think first continuistically for a while, then atomistically for a while. If we always thought in the same way, we would not be able to catch a breath of mental air. Thus our fatal uncertainty is confirmed and justified. Theories begin to look like arbitrary whims. We no longer live in any real connection with the world. We merely think of various ways in which we might live with the world, depending on our own subjective needs.

What would the old way of thought have said in such a case? It would have said: In an age when the leading thinkers think continuistically, they are thinking mainly of life. In one in which they think atomistically, they are thinking primarily of death, of inorganic nature, and they view even the organic in inorganic terms.

This is no longer unjustified arbitrariness. This rests on an objective relationship to things. Naturally, I can take turns in dealing with the animate and the inanimate. I can say that the very nature of the animate requires that I conceive of it continuistically, whereas the nature of the inanimate requires that I think of it atomistically. But I cannot say that this is only due to the arbitrary nature of the human mind. On the contrary, it corresponds to an objective relating of oneself to the world. For such perception, the subjective aspect is really disregarded, because one recognizes the animate in nature in continual form and the inanimate in discrete form. And if one really has to oscillate between the two forms of thought, this can be turned in an objective direction by saying that one approach is suited to the living and the other is suited to the dead. But there is no justification for making everything subjective as Poincaré does. Nor is the subjective valid for the way of perception that belonged to earlier times.

The gist of this is that in the phase of scientific thinking immediately preceding our own, there was a turn away from the animate to the inanimate; i.e., from continuity to atomism. This was entirely justified, if rightly understood. But, if we hope to objectively and truly find ourselves in the world, we must find a way out of the dead world of atomism, no matter how impressive it is as a theory. We must get back to our own nature and comprehend ourselves as living beings. Up to now, scientific development has tended in the direction of the inanimate, the atomistic. When, in the first part of the Nineteenth Century, this whole dreadful cell theory of Schleiden[50] and Schwann[51] made its appearance, it did not lead to continuity but to atomism. What is more, the scientific world scarcely admitted this, nor has it to this day realized that it should admit it since atomism harmonizes with the whole scientific methodology. We were not aware that by conceiving the organism as divided up into cells, we actually atomized it in our minds, which in fact signifies killing it. The truth of the matter is that any real idea of organisms has been lost to the atomistic approach.

This is what we can learn if we compare Goethe's views on organics with those of Schleiden or the later botanists. In Goethe we find living ideas that he actually experiences. The cell is alive, so the others are really dealing with something organic, but the way they think is just as though the cells were not alive but atoms. Of course, empirical research does not always follow everything to its logical conclusion, and this cannot be done in the case of the organic world. Our comprehension of the organic world is not much aided by the actual observations resulting from the cell theory. The non-atomistic somehow finds its way in, since we have to admit that the cells are alive. But it is typical of many of to-day's scientific discussions that the issues become confused and there is no real clarity of thought.

LECTURE VI

Dornach, January 1, 1923[52]

In my last lecture, I said that one root of the scientific world conception lay in the fact that John Locke and other thinkers of like mind distinguished between the primary and secondary qualities of things in the surrounding world. Locke called primary everything that pertains to shape, to geometrical and numerical characteristics, to motion and to size. From these he distinguished what he called the secondary qualities, such as color, sound, and warmth. He assigned the primary qualities to the things themselves, assuming that spatial corporeal things actually existed and possessed properties such as form, motion and geometrical qualities; and he further assumed that all secondary qualities such as color, sound, etc. are only effects on the human being. Only the primary qualities are supposed to be in the external things. Something out there has size, form, and motion, but is dark, silent, and cold. This produces some sort of effect that expresses itself in man's experiences of sound, color and warmth.

I have also pointed out how, in this scientific age, space became an abstraction in relation to the dimensions. Man was no longer aware that the three dimensions—up-down, right-left, front-back—were concretely experienced within himself. In the scientific age, he no longer took this reality of the three dimensions into consideration. As far as he was concerned, they arose in total abstraction. He no longer sought the intersecting point of the three dimensions where it is in fact experienced; namely, within man's own being. Instead, he looked for it somewhere in external space, wherever it might be. Thenceforth, this space framework of the three dimensions had an independent existence, but only an abstract thought-out one. This empty thought was no longer experienced as belonging to the external world as well as to man, whereas an earlier age experienced the three spatial dimensions in such a way that man knew he was experiencing them not only in himself but together with the nature of physical corporeality.

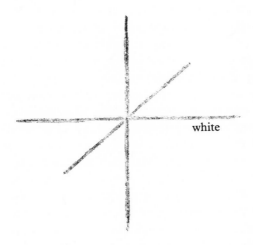

white

The dimensions of space had, as it were, already been abstracted and ejected from man. They had acquired a quite abstract, inanimate character. Man had forgotten that he experiences the dimensions of space in his own being together with the external world; and the same applied to everything concerned with geometry, number, weight, etc. He no longer knew that in order to experience them in their full living reality, he had to look into his own inner being. A man like John Locke transferred the primary qualities— which are of like kind with the three dimensions of space, the latter being a sort of form or shape—into the external world only because the connection of these qualities with man's inner being was no longer known.

The others, the secondary qualities, which were actually experienced qualitatively (as color, tone, warmth, smell, or taste), now were viewed as merely the effects of the things upon man, as inward experiences. But I have pointed out that inside the physical man as well as inside the etheric man these secondary qualities can no longer be found, so that they became free-floating in a certain respect. They were no longer sought in the outer world; they were relocated into man's inner being. It was felt that so long as man did not listen to the world, did not look at it, did not direct his sense of warmth to it, the world was silent. It had primary qualities, vibrations that were formed in a certain way, but no sound; it had processes of some kind in the ether, but no color; it had some sort of processes in ponderable matter (matter that has weight)—but it had no quality of warmth. As to these experienced qualities, the scientific age was really saying that it did not know what to do with them. It did not want to look for them in the world, admitting that it was powerless to do so. They were sought for within man, but only because nobody had any better idea. To a certain extent science investigates man's inner nature, but it does not (and perhaps cannot) go very far with this, hence it really does not take into consideration that these secondary qualities cannot be found in this inner nature. Therefore it has no pigeonhole for them. Why is this so?

Let us recall that if we really want to focus correctly on something that is related to form, space, geometry or arithmetic, we have to turn our attention to the inward life-filled activity whereby we build up the spatial element within our own organism, as we do with above-below, back-front, left-right. Therefore, we must say that if we want to discover the nature of geometry and space, if we want to get to the essence of Locke's primary qualities of corporeal things, we must look within ourselves. Otherwise, we only attain to abstractions.

In the case of the secondary qualities such as sound, color, warmth, smell, and taste, man has to remember that his ego and astral body normally dwell within his physical and etheric bodies but during sleep they can also be outside the physical and etheric bodies. Just as man experiences the primary qualities, such as the three dimensions, not outside but within himself during full wakefulness, so, when he succeeds (whether through instinct or through spiritual-scientific training) in really inwardly experiencing what is to be found outside the physical and etheric bodies from the moment of falling asleep to waking up, he knows that he is really experiencing the true essence of sound, color, smell, taste, and warmth in the external world outside his own body. When, during the waking condition, man is only within himself, he cannot experience anything but picture-images of the true realities of tone, color, warmth, smell and taste. But these images correspond to soul-spirit realities, not physical-etheric ones. In spite of the fact that what we experience as sound seems to be connected with certain forms of air vibrations, just as color is connected with certain processes in the colorless external world, it still has to be recognized that both are pictures, not of anything corporeal, but of the soul-spirit element contained in the external world.

We must be able to tell ourselves: When we experience a sound, a color, a degree of warmth, we experience an image of them. But we experience them as reality, when we are outside our physical body. We can portray the facts in a drawing as follows: Man experiences the primary qualities within himself when fully awake, and projects them as images into the outer world. If he only knows them in the outer world, he has the primary qualities only in images (arrow in sketch). These images are the mathematical, geometrical, and arithmetical qualities of things.

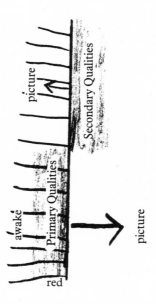

It is different in case of the secondary qualities. (The horizontal lines stand for the physical and etheric body of man, the red shaded area for the soul-spirit aspect, the ego and astral body.) Man experiences them outside his physical and etheric body,[53] and projects only the images into himself. Because the scientific age no longer saw through this, mathematical forms and numbers became something that man looked for abstractly in the outer world. The secondary qualities became something that man looked for only in himself. But because they are only images in himself, man lost them altogether as realities.

A few isolated thinkers, who still retained traditions of earlier views concerning the outer world, struggled to form conceptions that were truer to reality than those that, in the course of the scientific age, gradually emerged as the official views. Aside from Paracelsus,[54] there was, for example, van Helmont,[55] who was well aware that man's spiritual element is active when color, tone, and so forth are experienced. During the waking state, however, the spiritual is active only with the aid of the physical body. Hence it produces only an image of what is really contained in sound or color. This leads to a false description of external reality; namely, the purely mathematical-mechanistic form of motion for what is supposed to be experienced as secondary qualities in man's inner being, whereas, in accordance with their reality, their true nature, they can only be experienced outside the body. We should not be told that if we wish to comprehend the true nature of sound, for example, we ought to conduct physical experiments as to what happens in the air that carries to us the sound that we hear. Instead, we should be told that if we want to acquaint ourselves with the true nature of sound, we have to form an idea of how we really experience sound outside our physical and etheric bodies. But these are thoughts that never occurred to the men of the scientific age. They had no inclination to consider the totality of human nature, the true being of man. Therefore they did not find either mathematics or the primary qualities in this unknown human nature; and they did not find the secondary qualities in the external world, because they did not know that man belongs to it also.

I do not say that one has to be clairvoyant in order to gain the right insight into these matters, although a clairvoyant approach would certainly produce more penetrating perceptions in this area. But I do say that a healthy and open mind would lead one to place the primary qualities, everything mathematical-mechanical, into man's inner being, and to place the secondary qualities into the outer world. The thinkers no longer understood human nature. They did not know how man's corporeality is filled with spirit, or how this spirit, when it is awake in a person, must forget itself and devote itself to the body if it is to comprehend mathematics. Nor was it known that this same spirituality must take complete hold of itself and live independently of the body, outside the body, in order to come to the secondary qualities. Concerning all these matters, I say that clairvoyant perception can give greater insight, but it is not indispensable. A healthy and open mind can feel that mathematics belongs inside, while sound, color, etc. are something external.

In my notes on Goethe's scientific works[56] in the 1880's, I set forth what healthy feeling can do in this direction. I never mentioned clairvoyant knowledge, but I did show to what extent man can acknowledge the reality of color, tone, etc. without any clairvoyant perception. This has not yet been understood. The scientific age is still too deeply entangled in Locke's manner of thinking. I set it forth again, in philosophic terms, in 1911 at the Philosophic Congress in Bologna.[57] And again it was not understood. I tried to show how man's soul-spirit organization does indeed indwell and permeate the physical and etheric body during the waking state, but still remains inwardly independent. If one senses this inward independence of the soul and spirit, then one also has a feeling for what the soul and spirit have experienced during sleep about the reality of green and yellow, G and C-sharp, warm and cold, sour or sweet. But the scientific age was unwilling to go into a true knowledge of man.

This description of the primary and secondary qualities shows quite clearly how man got away from the correct feeling about himself and his connection to the world. The same thing comes out in other connections. Failing to grasp how the mathematical with its three-dimensional character dwells in man, the thinkers likewise could not understand man's spirituality. They would have had to see how man is

78

in a position to comprehend right-left by means of the symmetrical movements of his arms and hands and other symmetrical movements. Through sensing the course taken, for example, by his food, he can experience front-back. He experiences up-down as he coordinates himself in this direction in his earliest years. If we discern this, we see how man inwardly unfolds the activity that produces the three dimensions of space. Let me point out also that the animal does not have the vertical direction in the same way as man does, since its main axis is horizontal, which is what man can experience as front-back. The abstract space framework could no longer produce anything other than mathematical, mechanistic, abstract relationships in inorganic nature. It could not develop an inward awareness of space in the animal or in man.

Thus no correct opinion could be reached in this scientific age concerning the question: How does man relate to the animal, the animal to man? What distinguishes them from one another? It was still dimly felt that there was a difference between the two, hence one looked for the distinguishing features. But nothing could be found in either man or animal that was decisive and consistent. Here is a famous example: It was asserted that man's upper jawbone, in which the upper teeth are located, was in one piece, whereas in the animal, the front teeth were located in a separate bone, the inter-maxillary bone, with the actual upper jawbone on either side of them. Man, it was thought, did not possess this inter-maxillary bone. Since one could no longer find the relationship of man to animal by inner soul-spirit means, one looked for it in such external features and said that the animal had an inter-maxillary bone and man did not.

Goethe could not put into words what I have said today concerning primary and secondary qualities. But he had a healthy feeling about all these matters. He knew instinctively that the difference between man and animals must lie in the human form as a whole, not in any single feature. This is why Goethe opposed the idea that the inter-maxillary bone is missing in man. As a young man, he wrote an important article suggesting that there is an inter-maxillary bone in man as well as in the animal. He was able to prove this by showing that in the embryo the inter-maxillary bone is still clearly evident in man although in early childhood this bone fuses with the upper jaw, whereas it remains separate in the animal. Goethe did all this out of a certain instinct, and this instinct led him to say that one must not seek the difference between man and animal in details of this kind; instead, it must be sought for in the whole relation of man's form, soul, and spirit to the world.

By opposing the naturalists who held that man lacks the inter-maxillary bone Goethe brought man close to the animal. But he did this in order to bring out the true difference as regards man's essential nature. Goethe's approach out of instinctive knowledge put him in opposition to the views of orthodox science, and this opposition has remained to this day. This is why Goethe really found no successors in the scientific world. On the contrary, as a consequence of all that had developed since the Fifteenth Century in the scientific field, in the Nineteenth Century the tendency grew stronger to approximate man to the animal. The search for a difference in external details diminished with the increasing effort to equate man as nearly as possible with the animal. This tendency is reflected in what arose later on as the Darwinian idea of evolution. This found followers, while Goethe's conception did not. Some have treated Goethe as a kind of Darwinist, because all they see in him is that, through his work on the inter-maxillary bone,[58] he brought man nearer to the animal. But they fail to realize that he did this because he wanted to point out (he himself did not say so in so many words, but it is implicit in his work) that the difference between man and animal cannot be found in these external details.

Since one no longer knew anything about man, one searched for man's traits in the animal. The conclusion was that the animal traits are simply a little more developed in man. As time went by, there was no longer any inkling that even in regard to space man had a completely different position. Basically, all views of evolution that originated during the scientific age were formulated without any true knowledge of man. One did not know what to make of man, so he was simply represented as the culmination of the animal series. It was as though one said: Here are the animals; they build up to a final degree of perfection, a perfect animal; and this perfect animal is man.

My dear friends, I want to draw your attention to how matters have proceeded with a certain inner consistency in the various branches of scientific thinking since its first beginnings in the Fifteenth Century; how we picture our relation to the world on the basis of physics, of physiology, by saying: Out there is a silent and colorless world. It affects us. We fashion the colors and sounds in ourselves as experiences of the effects of the outer world.

At the same time we believe that the three dimensions of space exist outside of us in the external world. We do this, because we have lost the ability to comprehend man as a whole. We do this because our theories of animal and man do not penetrate the true nature of man. Therefore, in spite of its great achievements we can say that science owes its greatness to the fact that it has completely missed the essential nature of man. We were not really aware of the extent to which science was missing this. A few especially enthusiastic materialistic thinkers in the Nineteenth Century asserted that man cannot rightly lay claim to anything like soul and spirit because what appears as soul and spirit is only the effect of something taking place outside us in time and space. Such enthusiasts describe how light works on us; how something etheric (according to their theory) works into us through vibrations along our nerves; how the external air also continues on in breathing, etc. Summing it all up, they said that man is dependent on every rise and fall of temperature, on any malformation of his nervous system, etc. Their conclusion was that man is a creature pitifully dependent on every draft or change of pressure.

Anyone who reads such descriptions with an open mind will notice that, instead of dealing with the true nature of man, they are describing something that turns man into a nervous wreck. The right reply to such descriptions is that a man so dependent on every little draft of air is not a normal person but a neurasthenic. But they spoke of this neurasthenic as if he were typical. They left out his real nature, recognizing only what might make him into a neurasthenic. Through the peculiar character of this kind of thinking about nature, all understanding was gradually lost. This is what Goethe revolted against, though he was unable to express his insights in clearly formulated sentences.

Matters such as these must be seen as part of the great change in scientific thinking since the Fifteenth Century. Then they will throw light on what is essential in this development. I would like to put it like this: Goethe in his youth took a keen interest in what science had produced in its various domains. He studied it, he let it stimulate him, but he never agreed with everything that confronted him, because in all of it he sensed that man was left out of consideration. He had an intense feeling for man as a whole. This is why he revolted in a variety of areas against the scientific views that he saw around him. It is important to see this scientific development since the Fifteenth Century against the background of Goethe's world conception. Proceeding from a strictly historical standpoint, one can clearly perceive how the real being of man is missing in the scientific approach, missing in the physical sciences as well as in the biological.

This is a description of the scientific view, not a criticism. Let us assume that somebody says: "Here I have water. I cannot use it in this state. I separate the oxygen from hydrogen, because I need the hydrogen." He then proceeds to do so. If I then say what he has done, this is not criticism of his conduct. I have no business to tell him he is doing something wrong and should leave the water alone. Nor is it criticism, when I say that since the Fifteenth Century science has taken the world of living beings and separated from it the true nature of man, discarding it and retaining what this age required. It then led this dehumanized science to the triumphs that have been achieved.

It is not a criticism if something like this is said; it is only a description. The scientist of modern times needed a dehumanized nature, just as a chemist needs deoxygenized hydrogen and therefore has to split water into its two components. The point is to understand that we must not constantly fall into the error of looking to science for an understanding of man.

LECTURE VII

Dornach, January 2, 1923

Continuing with yesterday's considerations concerning the inability of the scientific world conception to grasp the nature of man, we can say that in all domains of science something is missing that is also absent in the mathematical-mechanistic sphere. This sphere has been divorced from man, as if man were absent from the mathematical experience. This line of thought results in a tendency to also separate other processes in the world from man. This in its turn produces an inability to create a real bridge between man and world.

I shall discuss another consequence of this inability later on. Let us focus first of all on the basic reason why science has developed in this way. It was because we lost the power to experience inwardly something that is spoken of in Anthroposophy today and that in former times was perceived by a sort of instinctive clairvoyance. Scientific perception has lost the ability to see into man and grasp how he is composed of different elements.

Let us recall the anthroposophical idea that man is composed of four members—the physical body, the etheric body, the astral body, and the I-organization. I need not go into detail about this formation, since you can find it all in my book *Theosophy*.[59] When we observe the physical body and consider the possibility of inward orientation—in other words, the possibility of inwardly experiencing one's physical body—we should begin by asking: What do we experience in regard to it? We experience what I have frequently spoken about recently; namely, the right-left, up-down, and front-back directions. We experience motion, the change of place of one's own body. To some extent at least, we also experience weight in various degrees. But weight is experienced in a highly modified form. When these things were still experienced within our various members, we reflected on them a good deal; but in the scientific age, no one gives them any thought. Facts that are of monumental importance for a world comprehension are completely ignored. Take the following fact. Assume that you have to carry a person who weighs as much as you do. Imagine that you carry this person a certain distance. You will consciously experience his weight. Of course, as you walk this distance, you are carrying yourself as well. But you do not experience this in the same way. You carry your own weight through space, but you do not experience this. Awareness of one's own weight is something quite different. In old age, we are apt to say that we feel the weight of our limbs. To some extent this is connected with weight, because old age entails a certain disintegration of the organism. This in turn tears the individual members out of the inward experience and makes them independent—atomizes them, as it were—and in atomization they fall a prey to gravity. But we do not actually feel this at any given moment of our life, so this statement that we feel the weight of our limbs is really only a figure of speech. A more exact science might show that it is not purely figurative, but be that as it may, the experience of our weight does not impinge strongly on our consciousness.

This shows that we have an inherent need to obliterate certain effects that are unquestionably working within us. We obliterate them by means of opposite effects ("opposite" in the sense brought out by the analogy between man and the course of the year in my recent morning lectures[60]). Nevertheless, whether we are dealing with processes that can be experienced relatively clearly, such as the three dimensions or motion, or with less obvious ones such as those connected with weight, they are all processes that can be experienced in the physical body.

What was thus experienced in former times has since been completely divorced from man. This is most evident in the case of mathematics. The reason it is less obvious in other experiences of the physical body is that the corresponding processes in the body, such as weight or gravity, are completely extinguished for today's form of consciousness. These processes, however, were not always completely obliterated. Under the influence of the mood prevailing under the scientific world conception, people today no longer have any idea of how different man's inner awareness was in the past. True, he did not consciously carry his weight through space in former times. Instead, he had the feeling that along with this weight, there was a counterweight. When he learned something, as was the case with the neophytes in the mysteries, he learned to perceive how, while he always carried his own weight in and with himself, the counter-effect is constantly active in light. It can really be said that man felt that he had to thank the spiritual element indwelling the light for counteracting, within him, the soul-spirit element active in gravity. In short, we can show in many ways that in older times there was no feeling that anything was completely divorced from man. Within himself, man experienced the processes and events as they occurred in nature. When he observed the fall of a stone, for example, in external nature (an event physically separated from him) he experienced the essence of movement. He experienced this by comparing it with what such a movement would be like in himself. When he saw a falling stone, he experienced something like this: "If I wanted to move in the same way, I would have to acquire a certain speed, and in a falling stone the speed differs from what I observe, for instance, in a slowly crawling creature." He experienced the speed of the falling stone by applying his experience of movement to the observation of the falling stone. The processes of the external world that we study in physics today were in fact also viewed objectively by the man of former times, but he gained his knowledge with the aid of his own experiences in order to rediscover in the external world the processes going on within himself.

Until the beginning of the Fifteenth Century, all the conceptions of physics were pervaded by something of which one can say that it brought even the physical activities of objects close to the inner life of man. Man experienced them in unison with nature. But with the onset of the Fifteenth Century begins the divorce of the observation of such processes from man. Along with it came the severance of mathematics, a way of thinking which from then on was combined with all science. The inner experience in the physical body was totally lost. What can be termed the inner physics of man was lost. External physics was divorced from man, along with mathematics. The progress thereby achieved consisted in the objectifying of the physical. What is physical can be looked at in two ways. Staying with the example of the falling stone, it can be traced with external vision. It can also be brought together with the experience of the speed that would have to be achieved if one wanted to run as fast as the stone falls. This produces comprehension that goes through the whole man, not one related only to visual perception.

To see what happened to the older world view at the dawn of the Fifteenth Century, let us look at a man in whom the transition can be observed particularly well; namely, Galileo.[61] Galileo is in a sense the discoverer of the laws governing falling objects. Galileo's main aim was to determine the distance travelled in the first second by a falling body. The older world view placed the visual observation of the falling stone side by side with the inward experience of the speed needed to run at an equal pace. The inner experience was placed alongside that of the falling stone. Galileo also observed the falling stone, but he did not compare it with the inward experience. Instead, he measured the distance travelled by the stone in the first second of its fall. Since the stone falls with increasing speed, Galileo also measured the following segments of its path. He did not align this with any inward experience, but with an externally measured process that had nothing to do with man, a process that was completely divorced from man. Thus, in perception and knowledge, the physical was so completely removed from man that he was not aware that he had the physical inside him as well.

At that time, around the beginning of the Seventeenth Century, a number of thinkers who wanted to be progressive began to revolt against Aristotle,[62] who throughout the Middle Ages had been considered the preeminent authority on science. If Aristotle's explanations of the falling stone (misunderstood in most cases today) are looked at soberly, we notice that when something is beheld in the world outside, he always points out how it would be if man himself were to undergo the same process. For him, it is not a matter of determining a given speed by measuring it, but to think of speed in such a way that it can be related to some human experience. Naturally, if you say you must achieve a particular speed, you feel that something alive, something filled with vigor, will be needed for you to do this. You feel a certain inner impetus, and the last thing you would assume is that something is pulling you in the direction you were heading. You would think that you were pushing, not that you were being pulled. This is why the force of attraction, gravity, begins to mean something only in the Seventeenth Century.

Man's ideas about nature began to change radically; not just the law of falling bodies but all the ideas of physics. Another example is the law of inertia, as it is generally called. The very name reveals its origin within man.* Inertia is something that can be inwardly felt but what has become of the law of inertia in physics under the influence of "Galileoism?" The physicist says: A body, or rather a point, on which no external influence is exercised, which is left to itself, moves through space with uniform velocity. This means that throughout all time-spans it travels the same distance in each second. If no external influence interferes, and the body has achieved a given speed per second, it travels the same distance in each succeeding second.

It is inert. Lacking an external influence, it continues on and on without change. All the physicist does is measure the distance per second, and a body is called inert if the velocity remains constant.

There was a time when one felt differently about this and asked: How is a moving body, traveling a constant distance per second, experienced? It could be experienced by remaining in one and the same condition without ever changing one's behavior. At most, this could only be an ideal for man. He can attain this ideal of inertia only to a very small degree. But if you look at what is called inertia in

*There is a play on words here. The German term for inertia, Trägheit, really means laziness.

ordinary life, you see that it is pretty much like doing the same thing every second of your life.

From the Fifteenth Century on, the whole orientation of the human mind was led to such a point that we can fairly say that man forgot his own inward experience. This happens first with the inner experience of the physical organism —man forgets it. What Galileo thought out and applied to matters close to man, such as the law of inertia, was now applied in a wide context. And it was indeed merely thought out, even if Galileo was dealing with things that can be observed in nature.

We know how, by placing the sun in the center instead of the earth, and by letting the planets move in circles around the sun, and by calculating the position of a given planetary body in the heavens, Copernicus produced a new cosmic system in a physical sense. This was the picture that Copernicus drew of our planetary, our solar system. And it was a picture that certainly can be drawn. Yet, this picture did not make a radical turn toward the mathematical attitude that completely divorces the external world from man. Anyone reading Copernicus's text gets the impression that Copernicus still felt the following. In the complicated lines, by means of which the earlier astronomy tried to grasp the solar system, it not only summed up the optical locations of the planets; it also had a feeling for what would be experienced if one stood amid these movements of the planets. In former ages people had a very clear idea of the epicycles the planets were thought to describe. In all this there was still a certain amount of human feeling. Just as you can understand the position of, let us say, an arm when you are painting a picture of a person because you can feel what it is like to be in such a position, so there was something alive in tracing the movement described by a planet around its fixed star. Indeed, even in Kepler's[63] case—perhaps especially in his case—there is still something of a human element in his calculating the orbits described by the planets.

Now Newton applies Galileo's abstracted principle to the heavenly bodies, adopting something like the Copernican view and conceiving things somewhat as follows: A central body, let us say a sun, attracts a planet in such a way that this force of attraction decreases in proportion to the square of the distance. It becomes smaller and smaller in proportion to the square, but increases in proportion to the mass of the bodies. If the attracting body has a greater mass, the force of attraction is proportionately greater.

If the distance is greater, the force of attraction decreases, but always in such a way that if the distance is twice as great, the attraction is four times less; if it is three times as great, nine times less, and so forth. Pure measuring is instilled into the picture, which, again, is conceived as completely abstracted from man. This was not yet so with Copernicus and Kepler but with Newton, a so-called "objective" something is excogitated and there is no longer any experience, it is all mere excogitation. Lines are drawn in the direction in which one looks and forces are, as it were, imagined into them, since what one sees is not force; the force has to be dreamed up. Naturally, one says "thought up" as long as one believes in the whole business; but when one no longer has faith in it, one says "dreamed up."

Thus we can say that through Newton the whole abstracted physical mode of conception becomes generalized so far that it is applied to the whole universe. In short, the aim is to completely forget all experience within man's physical body; to objectify what was formerly pictured as closely related to the experience of the physical body; to view it in outer space independent of the physical corporeality, although this space had first been torn out of the body experience; and to find ways to speak of space without even thinking about the human being. Through separation from the physical body, through separation of nature's phenomena from man's experience in the physical body, modern physics arises. It comes into existence along with this separation of certain processes of nature from self-experience within the physical human body (yellow in sketch). Self-experience is forgotten (red in sketch on page 5).

By permeating all external phenomena with abstract mathematics, this kind of physics could no longer understand man. What had been separated from man could not be reconnected. In short, there emerges a total inability to bring science back to man.

In physical respects you do not notice this quite so much; but you do notice it if you ask: What about man's self-experience in the etheric body, in this subtle organism? Man experiences quite a bit in it. But this was separated from man even earlier and more radically. This abstraction, however, was not as successful as in physics. Let us go back to a scientist of the first Christian centuries, the physician

Galen.[64] Looking at what lived in external nature and following the traditions of his time, Galen distinguished four elements—earth, water, air, and fire (we would say warmth). We see these if we look at nature. But, looking inward and focussing on the self-experience of the etheric body,[65] one asks: How do I experience these elements, the solid, the watery, the airy and the fiery, in myself? Then, in those times the answer was: I experience them with my etheric body. One experienced it as inwardly felt movements of the fluids: the earth as "black gall," the watery as "phlegm," the airy as "pneuma" (what is taken in through the breathing process), and warmth as "blood." In the fluids, in what circulates in the human organism, the same thing was experienced as what was observed externally. Just as the movement of the falling stone was accompanied by an experience in the physical body, so the elements were experienced in inward processes. The metabolic process, where (so it was thought) gall, phlegm, and blood work into each other, was felt as the inner experience of one's own body, but a form of inward experience to which corresponded the external processes occurring between air, water, fire, and earth.

Warmth – Blood   – Ego Organization
Air   – Pneuma   – Astral Body
Water   – Phlegm   – Etheric Body   – Chemistry
Earth   – Black Gall – Physical Body   – Physics

Here, however, we did not succeed in completely forgetting all inner life and still satisfying external observation. In the case of a falling body, one could measure something; for example, the distance travelled in the first second. One arrived at a "law of inertia" by thinking of moving points that do not alter their condition of movement but maintain their speed. By attempting to eject from the inward experience

93

something that the ancients strongly felt to be a specific inner experience; namely, the four elements, one was able to forget the inner content but one could not find in the external world any measuring system. Therefore the attempt to objectify what related to these matters, as was done in physics, remained basically unsuccessful to this day. Chemistry could have become a science that would rank alongside physics, if it had been possible to take as much of the etheric body into the external world as was accomplished in the physical body. In chemistry, however, unlike physics, we speak to this day of something rather undefined and vague, when referring to its laws.[66] What was done with physics in regard to the physical body was in fact the aim of chemistry in regard to the etheric body. Chemistry states that if substances combine chemically, and in doing so can completely alter their properties, something is naturally happening.

|||| yellow

//// red

But if one wants to go beyond this conception, which is certainly the simplest and most convenient, one really does not know much about this process. Water consists of hydrogen and oxygen; the two must be conceived as mixed together in the water somehow but no inwardly experiencable concept can be formed of this. It is commonly explained in a very external way: hydrogen consists of atoms (or molecules if you will) and so does oxygen. These intermingle, collide, and cling to one another, and so forth. This means that, although the inner experience was forgotten, one did not find oneself in the same position as in physics, where one could measure (and increasingly physics became a matter of

94

measuring, counting and weighing). Instead, one could only hypothesize the inner process. In a certain respect, it has remained this way in chemistry to this day, because what is pictured as the inner nature of chemical processes is basically only something read into them by thought.

Chemistry will attain the level of physics only when, with full insight into these matters, we can again relate chemistry with man, though not, of course, with the direct experience possessed by the old instinctive clairvoyance. We will only succeed in this when we gain enough insight into physics to be able to consolidate our isolated fragments of knowledge into a world conception and bring our thoughts concerning the individual phenomena into connection with man. What happens on one side, when we forget all inner experience and concentrate on measuring externals (thus remaining stuck in the so-called "objective") takes its revenge on the other side. It is easy enough to say that inertia is expressed by the movement of a point that travels the same distance in each succeeding second. But there is no such point. This uniform movement occurs nowhere in the domain of human observation. A moving object is always part of some relationship, and its velocity is hampered here or there. In short, what could be described as inert mass,[67] or could be reduced to the law of inertia, does not exist. If we speak of movement and cannot return to the living inner accompanying experience of it, if we cannot relate the velocity of a falling body to the way we ourselves would experience this movement, then we must indeed say that we are entirely outside the movement and must orient ourselves by the external world. If I observe a moving body (see sketch), and if these are its successive positions, I must somehow perceive that this body moves. If behind it there is a stationary wall, I follow the direction of movements and tell myself that the body moves on in that direction. But what is necessary in addition is that from my own position (dark circle) I guide this observation, in other words, become aware of an inward experience. If I completely leave out the human being and orient myself only out there, then, regardless of whether the object moves or remains stationary, while the wall moves, the result will be the same. I shall no longer be able to distinguish whether the body moves in one or the wall behind it in the opposite direction. I can basically make all the calculations under either one or the other assumption.

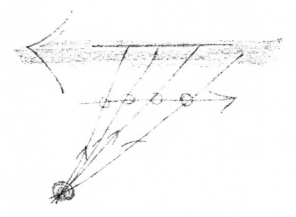

I lose the ability to understand a movement inwardly if I do not partake of it with my own experience. This applies, if I may say so, to many other aspects of physics. Having excluded the participating experience, I am prevented from building any kind of bridge to the objective process. If I myself am running, I certainly cannot claim that it is a matter of indifference whether I run or the ground beneath me moves in the opposite direction. But if I am watching another person moving over a given area, it makes no difference for merely external observation whether he is running or the ground beneath him is moving in the opposite direction. Our present age has actually reached the point, where we experience, if I may put it this way, the world spirit's revenge for our making everything physical abstract.

Newton was still quite certain that he could assume absolute movements, but now we can see numerous scientists trying to establish the fact that movement, the knowledge of movement, has been lost along with the inner experience of it. Such is the essence of the Theory of Relativity,[68] which is trying to pull the ground from under Newtonism. This theory of relativity is a natural historical result. It cannot help but exist today. We will not progress beyond it if we remain with those ideas that have been completely separated from the human element. If we want to understand rest or motion, we must partake in the experience. If we do not do this, then even rest and motion are only relative to one another.

LECTURE VIII

Dornach, January 3, 1923

I have tried to show how various domains of scientific thought originated in modern times. Now I will try to throw light from a certain standpoint on what was actually happening in the development of these scientific concepts. Then we shall better understand what these concepts signify in the whole evolutionary process of mankind. We must clearly understand that the phenomena of external culture are inwardly permeated by a kind of pulse beat that originates from deeper insights. Such insights need not always be ones that are commonly taught, but still they are at the bottom of the development. Now, I would only like to say that we can better understand what we are dealing with in this direction if we include in our considerations what in certain epochs was practiced as initiation science, a science of the deeper foundations of life and history.

We know that the farther we go back in history,[69] the more we discover an instinctive spiritual knowledge, an instinctive clairvoyant perception of what goes on behind the scenes. Moreover, we know that it is possible in our time to attain to a deeper knowledge, because since the last third of the Nineteenth Century, after the high tide of materialistic concepts and feelings, simply through the relationship of the spiritual world to the physical, the possibility arose to draw spiritual knowledge once again directly from the supersensible world. Since the last third of the Nineteenth Century, it has been possible to deepen human knowledge to the point where it can behold the foundations of what takes place in the external processes of nature.

So we can say that an ancient instinctive initiation science made way for an exoteric civilization in which little was felt of any direct spirit knowledge. Today, however, there is a new dawn of spirit knowledge, but now it is fully conscious rather than instinctive.

We stand at the beginning of this development of a new spirit knowledge. It will unfold further in the future. If we have a certain insight into what man regarded as knowledge during the age of the old instinctive science of initiation, we can discover that until the beginning of the Fourteenth Century, opinions prevailed in the civilized world that cannot be directly compared with any of our modern conceptions about nature. They were ideas of quite a different kind. Still less can they be compared with what today's science calls psychology. There too, we would have to say that it is of quite a different kind. The soul and spirit of man as well as the physical realm of nature were grasped in concepts and ideas that today are understood only be men who specifically study initiation science. The whole manner of thinking and feeling was quite different in former times.

If we examine the ancient initiation science, we find that, in spite of the fragmentary way in which it has been handed down, it had profound insights, deep conceptions, concerning man and his relation to the world.

People today do not greatly esteem a work like *De Divisione Naturae (Concerning the Division of Nature)* by John Scotus Erigena[70] in the Ninth Century. They do not bother with it because such a work is not regarded as an historical document since it comes from a time when men thought differently from the way they think today, so differently that we can no longer understand such a book. When ordinary philosophers describe such topics in their historical writings, one is offered mere empty words. Scholars no longer enter into the fundamental spirit of a work such as that of John Scotus Erigena on the division of nature, where even the term *nature* signifies something other than in modern science. If, with the insight of spiritual science, we do enter into the spirit of such a text, we must come to the following rather odd conclusion: This Scotus Erigena developed ideas that give the impression of extraordinary penetration into the essence of the world, but he presented these ideas in an inadequate and ineffective form. At the risk of speaking disrespectfully of a work that is after all very valuable, one has to say that Erigena himself no longer fully understood what he was writing about. One can see that in his description. Even for him, though not to the same extent as with modern historians of philosophy, the words that he had gleaned from tradition were more or less words only, and he could no longer enter into their deeper meaning. Reading his works, we find ourselves increasingly obliged to go farther back in history. Erigena's writings lead us directly back to those of the so-called pseudo-Dionysius the Areopagite.[71] I will now leave aside the historical problem of when Dionysius lived, and so forth. But again from Dionysius the Areopagite one is led still farther back. To continue the search one must be equipped with spiritual science. But finally, going back to the second and third millenia before Christ, one comes upon very deep insights that have been lost to mankind. Only as a faint echo are they present in writings such as those of Erigena.

Even if we go no further back than the Scholastics, we can find, hidden under their pedantic style, profound ideas concerning the way in which man apprehends the outer world, and how there lives the supersensible on one side and on the other side the sense perceptible, and so on. If we take the stream of tradition founded on Aristotle who, in his logical but pedantic way, had in turn gathered together the ancient knowledge that had been handed down to him, we find the same thing—deep insights that were well understood in ancient times and survived feebly into the Middle Ages, being repeated in the successive epochs, and were always less and less understood. That is the characteristic process. At last, in the Thirteenth or Fourteenth Century, the understanding disappears almost entirely, and a new spirit emerges, the spirit of Copernicus and Galileo, which I have described in the previous lectures.

In all studies, such as those I have just outlined, it is found that this ancient knowledge is handed down through the ages until the Fourteenth Century, though less and less understood. This ancient knowledge amounted essentially to an inner experience of what goes on in man himself. The explanations of the last few lectures should make this comprehensible: It is the experiencing of the mathematical-mechanical element in human movement, the experiencing of a certain chemical principle, as we would say today, in the circulation of man's bodily fluids, which are permeated by the etheric body. Hence, we can even look at the table that I put on the blackboard yesterday (page 93) from an historical standpoint. If we look at the being of man with our initiation science today, we have the physical body, the etheric body, the astral body (the inner life of the soul), and the ego organization. As I pointed out yesterday, there existed (arising out of the ancient initiation science) an inner experience of the physical body, an inward experience of movement, an inner experience of the dimensionality of space, as well as experiences of other physical and mechanical processes. We can call this inner experience the experiencing of physics in man. But this experience of physics in man is at the same time the cognition of the very laws of physics and mechanics. There was a physics of man directed towards the physical body. It would not have occurred to anyone in those times to search for physics other than through the experience in man. Now, in the age of Galileo and Copernicus, together with the mathematics that was thenceforth applied in physics, what was inwardly experienced is cast out of man and grasped abstractly. It can be said that physics sunders itself from man, whereas formerly it was contained in man himself.

Something similar was experienced with the fluid processes, the bodily fluids of the human organism. These too were inwardly experienced. Yesterday I referred to Galen who, in the first Christian centuries, described the following fluids in man: black gall, blood, phlegm, and the ordinary white or yellow gall. In the physical world man operates by means of the intermingling of these fluids by the way they influence each other. Galen did not arrive at these statements by anything resembling today's physiological methods. They were based mainly on inward experiences. For Galen too these were largely a tradition, but what he thus took from tradition was once experienced inwardly in the fluid part of the human organism, which in turn was permeated by the etheric body.

For this reason, in the beginning of my *Riddles of Philosophy*,[72] I did not describe the Greek philosophers in the customary way. Read any ordinary history of philosophy and you will find this subject presented more or less as follows: Thales[73] pondered on the origin of our sense world and sought for it in water. Heraclitus looked for it in fire. Others looked for it in air. Still others in solid matter, for example in something like atoms. It is amazing that this can be recounted without questions being raised. People today do not notice that it basically defies explanation why Thales happened to designate water while Heraclitus[74] chose fire as the source of all things. Read my book *The Riddles of Philosophy*, and you will see that the viewpoint of Thales, expressed in the sentence "All things have originated from water," is based on an inner experience. He inwardly felt the activity of what in his day was termed the watery element. He sensed that the basis of the external process in nature was related to this inner activity; thus he described the external out of inner experiences. It was the same with Heraclitus who, as we would say today, was of a different temperament. Thales, as a phlegmatic, was sensitive to the inward "water" or "phlegm." Therefore he described the world from the phlegmatic's viewpoint: everything has come from water. Heraclitus, as a choleric, experienced the inner "fire." He described the world the way he experienced it. Besides them, there were other thinkers, who are no longer mentioned by external tradition, who knew still more concerning these matters. Their knowledge was handed down and still existed as tradition in the first Christian centuries; hence Galen could speak of the four components of man's inner fluidic system.

What was then known concerning the inner fluids, namely, how these four fluids—yellow gall, black gall, blood, and phlegm—influence and mix with one another really amounts to an inner human chemistry, though it is of course considered childish today. No other form of chemistry existed in those days. The external phenomena that today belong to the field of chemistry were then evaluated according to these inward experiences. We can therefore speak of an inner chemistry based on experiences of the fluid man who is permeated by the ether body. Chemistry was tied to man in former ages. Later it emerged, as did mathematics and physics, and became external chemistry (see table). Try to imagine how the physics and chemistry of ancient times were felt by men. They were experienced as something that was, as it were, a part of themselves, not as something that is mere description of an external nature and its processes. The main point was: it was experienced physics, experienced chemistry.

In those ages when men felt external nature in their physical and etheric bodies, the contents of the astral body and the ego organization were also experienced differently than in later times.

Today, we have a psychology, but it is only an inventory

103

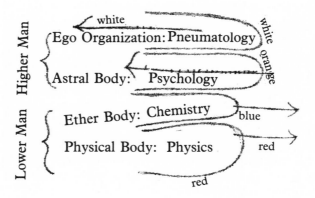

of abstractions, though no one admits this. You find in it thinking, feeling, willing, as well as memory, imagination, and so forth, but treated as mere abstractions. This gradually arose from what was still considered as one's own soul contents. One had cast out chemistry and physics; thinking, feeling and willing were retained. But what was left eventually became so diluted that it turned into no more than an inventory of lifeless empty abstractions, and it can be readily proved that this is so. Take, for example, the people who still spoke of thinking or willing as late as the Fifteenth or Sixteenth Century.[75] If you study the older texts on these subjects you will see that people expressed themselves concerning these matters in a concrete way. You have the feeling, when such a person speaks about thinking, that he speaks as if this thinking were actually a series of inner processes within him, as if the thoughts were colliding with each other or supporting each other. This is still an experiencing of thoughts. It is not yet as abstract a matter as it became later on. During and towards the end of the Nineteenth Century, it was an easy thing for the philosophers to deny all reality to these abstractions. They saw thoughts as inner mirror pictures, as was done in an especially brilliant way by Richard Wahle, who declared that the ego, thinking, feeling and willing were only illusions. Instead of abstractions, the inner soul contents become illusions.

In the age when man felt that his walking was a process that took place simultaneously in him and the world, and when he still sensed the circulating fluids within him, he knew, for instance, that when he moved about in the heat of the sun (when external influences were present) that the blood and phlegm circulated differently in him than was the case in winter. Such a man experienced the blood and phlegm circulation within himself, but he experienced it together with the sunshine or the lack thereof. And just as he experienced physical and chemical aspects in union with the outside world, so he also experienced thinking, feeling, and willing together with the world. He did not think they were occurring only within himself as was done in later ages when they gradually evaporated into complete abstractions. Instead he experienced what occurred in him in thinking, feeling, and willing, or in the circulation of the fluids as part of the realm of the astral, the soul being of man, which in that age was the subject of a psychology.

Psychology now became tightly tied to man. With the dawn of the scientific age, man drove physics and chemistry out into the external world; psychology, on the other hand, he drove into himself. This process can be traced in Francis Bacon and John Locke. All that is experienced of the external world, such as tone, color, and warmth, is pressed into man's interior.

This process is even more pronounced in regard to the ego organization. This gradually became a very meager experience. The way man looked into himself, the ego became by degrees something like a mere point. For that reason it became easy for philosophers to dispute its very existence. Not ego consciousness, but the experience of the ego was for men of former ages something rich in content and fully real.

This ego experience expressed itself in something that was a loftier science than psychology, a science that can be called pneumatology. In later times this was also pressed into the interior and thinned out into our present quite diluted ego feeling.

When man had the inward experience of his physical body, he had the experience of physics; simultaneously, he experienced what corresponds in outer nature to the processes in his physical body. It is similar in the case of the etheric body. Not only the etheric was experienced inwardly, but also the physical fluid system, which is controlled by the etheric. Now, what is inwardly experienced when man perceives the psychological, the processes of his astral body? The "air man"—if I may put it this way—is inwardly experienced. We are not only solid organic formations, not only fluids or watery formations, we are always gaseous-airy as well. We breathe in the air and breathe it out again. We experienced the substance of psychology in intimate union with the inner assimilation of air. This is why psychology was more concrete. When the living experience of air (which can also be outwardly traced) was cast out of the thought contents, these thought contents became increasingly abstract, became mere thought. Just think how an old Indian philosopher strove in his exercises to become conscious of the fact that in the breathing process something akin to the thought process was taking place. He regulated his breathing process in order to progress in his thinking. He knew that thinking, feeling, and willing are not as flimsy as we today make them out to be. He knew that through breathing they were related to both outer and inner nature, hence with air. As we can say that man expelled the physical and chemical aspects from his organization, we can also say that he sucked in the psychological aspect, but in doing so he rejected the external element, the air-breath experience. He withdrew his own being from the physical and chemical

106

elements and merely observed the outer world with physics and chemistry; whereas he squeezed external nature (air) out of the psychological. Likewise, he squeezed the warmth element out of the pneumatological realm, thus reducing it to the rarity of the ego.

If I call the physical and etheric bodies the "lower man" and call the astral body and ego-organization the "upper man," I can say that, in the transition from an older epoch to the scientific age, man lost the inner physical and chemical experience, and came to grasp external nature only with his concepts of physics and chemistry. In psychology and pneumatology, on the other hand, man developed conceptions from which he eliminated outer nature and came to experience only so much of nature as remained in his concepts. In psychology, this was enough so that he at least still had words for what went on in his soul. As to the ego, however, this was so little that pneumatology (partially because theological dogmatism had prepared this development) completely faded out. It shrank down to the mere dot of the ego.

All this took the place of what had been experienced as a unity, when men of old said: We have four elements, earth, water, air and fire. Earth we experience in ourselves when we experience the physical body. Water we experience in ourselves when we experience the etheric body as the agent that moves, mixes, and separates the fluids. Air is experienced when the astral body is experienced in thinking, feeling, and willing, because these three are experienced as surging with the inner breathing process.—Finally, warmth (or fire, as it was then called) was experienced in the sensation of the ego.

So we can say that the modern scientific view developed by way of a transformation of man's whole relation to himself. If you follow historical evolution with these insights, you will find what I told you earlier—that in each new epoch we see new descriptions of the old traditions, but these are always less and less understood. The works of men like Paracelsus, van Helmont, or Jacob Boehme,[76] bear witness to such ancient traditions.

One who has insight into these matters gets the impression that in Jacob Boehme's case a very simple man is speaking out of sources that would lead too far today to discuss. He is difficult to comprehend because of his clumsiness. But Jacob Boehme shows profound insight in his awkward descriptions, insights that have been handed down through the generations. What was the situation of a person like Jacob Boehme? Giordano Bruno, his contemporary, stood among the most advanced men of his time, whereas we see in Jacob Boehme's case that he obviously read all kinds of books that are naturally forgotten today. These were full of rubbish. But Boehme was able to find a meaning in them. Awkwardly and with great difficulty Boehme presents the primeval wisdom that he had gleaned from his still more awkward and inadequate sources. His inward enlightenment enabled him to return to an earlier stage.

If we now look at the Fifteenth, Sixteenth, and especially the Seventeenth and Eighteenth centuries, and if we leave aside isolated people like Paracelsus and Boehme (who appear like monuments to a bygone age), and if we look at the exoteric stream of human development in the light of initiation science, we gain the impression that nobody knows anything at all anymore about the deeper foundations of things. Physics and chemistry have been eliminated from man, and alchemy has become the subject of derision. Of course, people were justified in scoffing at it, because what still remained of the ancient traditions in medieval alchemy could well be made fun of. All that is left is psychology, which has become confined to man's inner being, and a very meager pneumatology. People have broken with everything that was formerly known of human nature. On one hand, they experience what has been separated from man; and on the other, what has been chaotically relegated into his interior. And in all our search for knowledge, we see what I have just described.

In the Seventeenth Century, a theory arose that remains quite unintelligible if considered by itself, although if it is viewed in the context of history it becomes comprehensible. The theory was that those processes in the human body that have to do with the intake of food, are based on a kind of fermentation. The foods man eats are permeated with saliva and then with digestive fluids such as those in the pancreas, and thus various degrees of fermentation processes, as they were called, are achieved. If one looks at these ideas from today's viewpoint (which naturally will also be outgrown in the future) one can only make fun of them. But if we enter into these ideas and examine them closely, we discover the source of these apparently foolish ideas. The ancient traditions, which in a man like Galen were based on inward experiences and were thus well justified, were now on the verge of extinction. At the same time, what was to become external objective chemistry was only in its beginnings. Men had lost the inner knowledge, and the external had not yet developed. Therefore, they found themselves able to speak about digestion only in quite feeble neo-chemical terms, such as the vague idea of fermentation. Such men were the late followers of Galen's teachings. They still felt that in order to comprehend man, one must start from the movements of man's fluids, his fluid nature. But at the same time, they were beginning to view chemical aspects only by means of the external processes. Therefore they seized the idea of fermentation, which could be observed externally, and applied it to man. Man had become an empty bag because he no longer experienced anything within himself. What had grown to be external science was poured into this bag. In the Seventeenth Century, of course, there was not much science to pour. People had vague ideas about fermentation and similar processes, and these were rashly applied to man. Thus arose the so-called iatrochemical school[77] of medicine.

In considering these iatrochemists, we must realize that they still had some inkling of the ancient doctrine of fluids, which was based on inner experience. Others, who were more or less contemporaries of the iatrochemists, no longer had any such inkling, so they began to view man the way he appears to us today when we open an anatomy book. In such books we find descriptions of the bones, the stomach, the liver, etc. and we are apt to get the impression that this is all there is to know about man and that he consists of more or less solid organs with sharply defined contours. Of course, from a certain aspect, they do exist. But the solid aspect—the earth element, to use the ancient terminology—comprises at most one tenth of man's organization. It is more accurate to say that man is a column of fluids. The mistake is not in what is actually said, but in the whole method of presentation. It is gradually forgotten that man is a column of fluids in which the clearly contoured organs swim. Laymen see the pictures and have the impression that this is all they need to understand the body. But this is misleading. It is only one tenth of man. The remainder ought to be described by drawing a continuous stream of fluids (see diagram) interacting in the most manifold ways in the stomach, liver, and so forth. Quite erroneous conceptions arise as to how man's organism actually functions, because only the sharply outlined organs are observed. This is why, in the Nineteenth Century, people were astonished to see that if one drinks a glass of water, it appears to completely penetrate the body and be assimilated by his organs. But when a second or third glass of water is consumed, it no longer gives the impression that it is digested in the same manner. These matters were noticed but could no longer be explained, because a completely false view was held con-

110

cerning the fluid organization of man. Here the etheric body is the driving agent that mixes or separates the fluids, and brings about the processes of organic chemistry in man.

In the Seventeenth Century, people really began to totally ignore this "fluid man" and to focus only on the solidly contoured parts. As time went by, man came to be regarded as a system of solid parts. In this realm of clearly outlined parts, everything takes place in a mechanical way. One part pushes another; the other moves; things get pumped; it all works like suction or pressure pumps. The body is viewed from a mechanical standpoint, as existing only through the

111

interplay of solidly contoured organs. Out of the iatrochemical theory or alongside it, there arose iatromechanics and even iatromathematics.[78]

Naturally, people began to think that the heart is really a pump that mechanically pumps the blood through the body, because they no longer knew that our inner fluids have their own life and therefore move on their own. They never dreamed that the heart is only a sense organ that checks on the circulation of the fluids in its own way. The whole matter was inverted. One no longer saw the movement and inner vitality of the fluids, or the etheric body active therein. The heart became a mechanical apparatus and has remained so to this day for the majority of physiologists and medical men.

The iatrochemists still had some faint knowledge concerning the etheric body. There was full awareness of it in what Galen described. In van Helmont or Paracelsus there was still an inkling of the etheric body, more than survived in the official iatrochemists who conducted the schools of that time. In the iatromechanists no trace whatsoever remained of this ether body; all conception of it had vanished into thin air. Man was seen only as a physical body, and that only to the extent that he consists of solid parts. These were now dealt with by means of physics, which had in the meantime also been cast out of the human being. Physics was now applied externally to man, whom one no longer understood. Man had been turned into an empty bag, and physics had been established in an abstract manner. Now this same physics was reapplied to man. Thus one no longer had the living being of man, only an empty bag stuffed with theories.

It is still this way today. What modern physiology or anatomy tells us of man is not man at all, it is physics that was cast out of man and is now changed around to be fitted back into man. The more intimately we study this development, the better we see destiny at work. The iatrochemists had a shadowy consciousness of the etheric body, the iatro-

112

mechanists had none. Then came a man by the name of Stahl[79] who, considering his time, was an unusually clever man. He had studied iatrochemistry, but the concepts of the "inner fermentation processes" seemed inadequate to him because they only transplanted externalized chemistry back into the human bag. With the iatromechanists he was still more dissatisfied because they only placed external mechanical physics back into the empty bag. No knowledge, no tradition existed concerning the etheric body as the driving force of the moving fluids. It was not possible to gain information about it. So what did Stahl do? He invented something, because there was nothing left in tradition. He told himself: the physical and chemical processes that go on in the human body cannot be based on the physics and chemistry that are discovered in the external world. But he had nothing else to put into man. Therefore he invented what he called the "life force," the "vital force." With it he founded the dynamic school. Stahl was gifted with a certain instinct. He felt the lack of something that he needed; so he invented this "vital force." The Nineteenth Century had great difficulty in getting rid of this concept. It was really only an invention, but it was very hard to rid science of this "life force."

Great efforts were made to find something that would fit into this empty bag that was man. This is why men came to think of the world of machines. They knew how a machine moves and responds. So the machine was stuffed into the empty bag in the form of *L'homme machine* by La Mettrie.[80] Man is a machine. The materialism, or rather the mechanics, of the Eighteenth Century, such as we see in Holbach's[81] *Systeme de la nature*, which Goethe so detested in his youth, reflects the total inability to grasp the being of man with the ideas that prevailed at that time in outer nature. The whole Nineteenth Century suffered from the inability to take hold of man himself.

But there was strong desire somehow or other to work

out a conception of man. This led to the idea of picturing him as a more highly evolved animal. Of course, the animal was not really understood either, since physics, chemistry, and psychology, all in the old sense, are needed for this purpose even if pneumatology is unnecessary. But nobody realized that all this is also required in order to understand the the animal. One had to start somewhere, so in the Eighteenth Century man was compared to the machine and in the Nineteenth Century he was traced back to the beast. All this is quite understandable from the historical standpoint. It makes good sense considering the whole course of human evolution. It was, after all, this ignorance concerning the being of man that produced our modern opinions about man. The development towards freedom, for example, would never have occurred had the ancient experience of physics, chemistry, psychology, and pneumatology survived. Man had to lose himself as an elemental being in order to find himself as a free being. He could only do this by withdrawing from himself for a while and paying no attention to himself any longer. Instead, he occupied himself with the external world, and if he wanted theories concerning his own nature, he applied to himself what was well suited for a comprehension of the outer world. During this interim, when man took the time to develop something like the feeling of freedom, he worked out the concepts of science; these concepts that are, in a manner of speaking, so robust that they can grasp outer nature. Unfortunately, however, they are too coarse for the being of man, since people do not go to the trouble of refining these ideas to the point where they can also grasp the nature of man. Thus modern science arose, which is well applicable to nature and has achieved great triumphs. But it is useless when it comes to the essential being of man.

You can see that I am not criticizing science. I am only

describing it. Man attains his consciousness of freedom only because he is no longer burdened with the insights that he carried within himself and that weighed him down. The experience of freedom came about when man constructed a science that in its robustness was only suited to outer nature. Since it does not offer the whole picture and is not applicable to man's being, this science can naturally be criticized in turn. It is most useful in physics; in chemistry, weak points begin to show up; and psychology becomes completely abstract. Nevertheless, mankind had to pass through an age that took its course in this way in order to attain to an individually modulated moral conception of the world and to the consciousness of freedom. We cannot understand the origin of science if we look at it only from one side. It must be regarded as a phenomenon parallel to the consciousness of freedom that is arising during the same period, along with all the moral and religious implications connected with this awareness.

This is why people like Hobbes[82] and Bacon, who were establishing the ideas of science, found it impossible to connect man to the spirit and soul of the universe. In Hobbes' case, the result was that, on the one hand, he cultivated the germinal scientific concepts in the most radical way, while, on the other hand, he cast all spiritual elements out of social life and decreed "the war of all against all." He recognized no binding principle that might flow into social life from a supersensible source, and therefore he was able, though in a somewhat caricatured form, to discuss the consciousness of freedom in a theoretical way for the first time.

The evolution of mankind does not proceed in a straight line. We must study the various streams that run side by side. Only then can we understand the significance of man's historical development.

LECTURE IX

Dornach, January 6, 1923

It is in the nature of the case that the subject of a lecture course like this one is inexhaustible. Matters could be elaborated and looked at more thoroughly. But since, unfortunately, we must come to an end, we have to be content with given guidelines and indications. Today, therefore, I shall only supplement the scanty outlines and hints already discussed so that in a certain sense the picture will be rounded out.

Proceeding once again from the being of man as viewed by spiritual science, we must say that we member man into physical body, etheric or formative forces body, astral body (which essentially represents the soul life) and ego. Let us be clear that properly speaking the physical body resides only in the small part of the human organization that we can describe as solid and sharply defined. On the other hand, all that pertains to liquid or fluid forms is taken hold of by the etheric body in such a way that it is in a constant process of blending, separating, combining, and dissolving. It is in perpetual flux. Then there are the gaseous, aeriform elements, such as are active in oxygen and other gases. In these, the astral body is at work. Finally, the ego organization is active in everything that has to do with warmth.

What I have just outlined cannot, however, be reduced to a diagram. We must clearly understand, for instance, that because the formative forces body pulsates through all fluid and liquid elements of the body, it also sweeps along the solid substances. Everything in the human organization is in close interaction, in constant interplay. We must always be

116

aware of that. But now let us also remember that this human organization has been experienced in different ways in the course of evolution. This was one of the main themes of these lectures.

What is described today as the subject matter of external physics or mechanics, was originally attained through an inward experience of the physical body. Our present-day physics contains statements that originated because there once existed an internally experienced physics of the physical body. As I have explained a number of times, this inward physics was divorced from man and now continues to function merely as a science that observes outer nature. During the decline of the medieval alchemy the same thing happened with what lives inwardly in man by virtue of the etheric body. The work of this body in the fluids was once experienced, but now it is only dimly perceptible in the fantastic, alchemistic formulas that we find in ancient writings. Originally this was intelligent science, but inwardly experienced within the etheric. In a way, this is still in the process of being divorced from man, because as yet we really do not have a fully developed chemistry. We have many chemical processes in the world that we seek to understand, but only in a physical and mechanical way.

In the beginning man experienced all this inwardly by means of his organization, but in the course of time he cast it all out of himself. In this process of casting out all our science developed, from astronomy to the meager beginnings of modern chemistry. On the other hand, thinking, feeling and willing, the subject matter of abstract psychology (which today is no longer considered real) was in former times actually not experienced inside man. Man felt himself at one with the external world outside his own being, when he experienced the soul life. Thus what was corporeal was once experienced inwardly, whereas the soul element was experienced by leaving one's being and communing with

117

the outer world. Psychology was once the science of that aspect of the world that affects man in such a way that he appears to himself as a soul being. Physics and chemistry were cast out of man, whereas psychology and pneumatology (which I shall discuss directly) were stuffed into him and lost their reality. They turned into subjective perceptions with which nothing could be done.

What was experienced together with the cosmos through the astral body (which leaves us in sleep) has become the subject of psychology. What man experienced as spirit in union with the universe was pneumatology. Today, as I have already pointed out, this has shrunk down to the idea of the ego or to a mere feeling. Therefore we now have as science of external nature what was once inner experience, while our science of man's inner nature is what was once external experience.

Now we must call to mind what is needed, on the one hand for physics and chemistry, and on the other for psychology and pneumatology, in order to develop them further in a conscious way, since man today finds himself in the age of the development of the consciousness soul. Take physics, for example, which in recent times has become mostly abstract and mechanical. From all that I have said you will have seen that the scientific age has increasingly felt impelled to restrict itself to the externally observed mechanics of space. Long ago, man accompanied motion by means of inward experience and judged it according to what he felt within as movement. Observing a falling stone, he experienced its inner impulse of movement in his own inner human nature, in his physical body. This experience, after the great casting out, led to the measuring of the rate of fall per second. In our attitude toward nature, the idea prevails that what is observed is what is real.

What can be observed in the outer world? It is motion, change of position.[83] As a rule, we let velocity vanish neatly in a differential coefficient. But it is motion that we observe, and we express velocity as movement per second, hence by means of space. This means, however, that with our conscious experience, we are entirely outside the object. We are not involved in it in any way when we merely watch its motion, meaning its change of position in space. We can do that only if we find ways and means to inwardly take hold of the spatial, physical object by an extending of the same method with which we separated from it in the first place. Instead of the mere movement, the bare change of position, we have to view the velocity in the objects as their characteristic element. Then we can know what a particular object is like inwardly, because we find velocity also within ourselves when we look back upon ourselves.

This is what is necessary. The trend of scientific development in regard to the outer physical world must be extended in the direction of proceeding from mere observation of motion to a feeling for the velocity possessed by a given object. We must advance from motion to velocity. That is how we enter into reality. Reality is not taken hold of if all we see is that a body changes its position in space. But if we know that the body possesses an inner velocity-impulse, then we have something that lies in the nature of the body. We assert nothing about a body if we merely indicate its change of position, but we do state something about it when we say that it contains within itself the impulse for its own velocity. This then is a property of it, something that belongs to its nature. You can understand this by a simple illustration. If you watch a moving person, you know nothing about him. But if you know that he has a strong urge to move quickly, you do know something about him. Likewise, you know something about him, when you know that he has a reason for moving slowly. We must be able to take hold of something that has significance *within* a given body. It matters little whether or not modern physics speaks, for example, of atoms; what matters is that when it does speak of them it regards them as velocity charges. That is what counts.

Now the question is: how do we arrive at such a perception? We can discuss this best in the case of physics, since today's chemistry has advanced too little. We have to become clear about what we actually do when, in our thinking, we cast inwardly experienced mechanics and physics into external space. That is what we are doing when we say: The nature of what is out there in space is of no concern to me; I observe only what can be measured and expressed in mechanical formulas, and I leave aside everything that is not mechanical. Where does this lead us? It leads us to the same process in knowledge that a human being goes through when he dies. When he dies, life goes out of him, the dead organism remains. When I begin to think mechanistically, life goes out of my knowledge. I then have a science of dead matter. We must be absolutely clear that we are setting up a science of dead matter so long as the mechanical and physical aspect is the sole object of our study of nature. You must be aware that you are focussing on what is dead. You must be able to say to yourself: The great thing about science is that it has tacitly resolved that, unlike the ancient alchemists who still saw in outer nature a remnant of life, it will observe what is dead in minerals, plants, and animals. Science will study only what is dead in them, because it utilizes only ideas and concepts suitable for what is dead. Therefore, our physics is dead by its nature.

Science will stand on a solid basis only when it fully realizes that its mode of thinking can take hold only of the dead. The same is true of chemistry, but I cannot go into that today because of the lack of time.

When we look only at motion and lose sight of velocity, we are erecting a physics that is dead, the end-product of living things is then our concern, and the end-product is death. Hence, when we look at nature with the eyes of modern mechanics and physics, we must realize that we are looking at a corpse.

Nature was not always like this. It was different at one time. If I look at a corpse, it would be foolish to believe that it was always in this condition. The fact that I realize that it is a corpse proves to me that once it was a living organism. The moment you realize that modern mechanics and physics lead you to view nature in this way, you will see that nature is now a corpse so far as physics is concerned. We are studying a corpse.

Can we attain to something living, or at least an approach to it? The corpse is the final condition of something living. Where is the beginning condition? Well, my dear friends, there is no way to rediscover velocity by observing motion. You may stare at differential coefficients as long as you will but you will not find it. Instead, you must turn back to man. Whereas formerly he experienced himself from within, you must now study him from without through his physical organism, and you must understand that in man—and especially in his physical and etheric organizations—the beginning of a living condition must be sought.

No satisfactory form of physics and chemistry will be attained save through a genuine science of man. But I expressly call attention to the fact that such a genuine anthropology will not be reached by approaching man with the methods of present-day physics and chemistry. That would only carry death back into man and make his body (his lower organization) even more dead than before.

You must study what is living in man, and not revert to the method of physics and chemistry. What is needed are the methods that can be found through spiritual-scientific research. Briefly stated, spiritual-scientific research will meet the historic requirements of natural science.

This historic requirement can be put in the following

words: Science has reached the point of observing what is corpse-like in nature. Anthroposophical spiritual science must discover in addition to this the beginning of a living condition. This has been preserved in man. In former periods of evolution it was also externally perceptible. At one time, the processes of nature were totally different. Today, we walk around on the corpses of what existed in the beginning. But in the two lower bodies of man, the beginning condition has been preserved. There we can discover all that once existed, right back to the Saturn condition. An historical approach leads beyond the present state of science. It is quite clear why this is so. We are in the midst of a period of development. If, as is so frequently the case, we consider today's manner of thinking to be the most advanced and do not realize that the real course of events was very different, then we are looking at history the wrong way. As an example, a twenty-five year old person need not only be observed in the light of the twenty-five years that he has been alive,—one must also observe the element in him that makes it possible for him to live on. That is one point.

Movement: *Velocity:* Dead Aspect (Final Condition of Being)
Phenomenon: *Being:*   Semblance (Initial Condition of Being)

The other point is that our psychology has become very thin, while pneumatology has nearly reached the vanishing point. Again, we must know how far it has gone with these two sciences in the present age. If one speaks today of blue or red, of C-sharp or G, or of qualities of warmth, he will say that they are subjective sensations. That is the popular attitude. But what is a mere subjective sensation? It is a "phenomenon." Just as we observe only motions in outer nature, we study only the phenomenon in psychology and pneumatology. And just as velocity is missing from motion in our external observation, the essential thing—the living essence

122

—is missing from our observation of the inner soul life. Because we only study phenomena and no longer experience the living essence, we never get beyond mere semblance. The way thinking, feeling, and willing are experienced today, they are mere semblance. Modern epistemologists have a dreadful time chewing on this semblance. They are like the man who wants to lift himself up by his own pigtail, or like the man in a railroad car who pushes against the wall without realizing that he cannot move the carriage in this way. This is how modern epistemologists look. They talk and talk, but there is no vitality in their talk because they are locked into the mere semblance.

I have tried to put a certain end to this talk. The first time was in my *Philosophy of Freedom*,[84] where I demonstrated how this semblance, inherent in pure thinking, becomes the impulse of freedom when inwardly grasped by man in thinking. If something other than semblance were contained in our subjective experience, we could never be free. But if this semblance can be raised to pure thinking, one can be free, because what is not real being cannot determine us, whereas real being would do so. This was my first effort. My second effort was at the Philosophical Congress in Bologna, when I analyzed the matter psychologically. I attempted to show that our sensations and thoughts are in fact outward experiences, rather than inward ones, and that this insight can be attained by careful observation.

These indications will have to be understood. Then, we shall realize that we must rediscover being in semblance, just as we must rediscover velocity in movement. Then, we will understand what this inwardly experienced semblance really is. It will reveal itself as the initial state of being. Man experiences this semblance; experiences himself as semblance and as such lives his way into semblance and thus transforms it into the seed of future worlds. I have often pointed out that from our ethics, our morals, born of the physical world of semblance, future physical worlds will arise, just as from today's seed the plant will grow.[85] We are dealing with the nascent state of being. In order to have a proper natural science, we must realize that psychology and pneumatology must understand what they observe as nascent states of being. Only then will they throw light on those matters that natural science wants to illuminate. But what is this "nascent" or "initial state?"

Now this nascent state is in the outer world, not within. It is in what I see when I behold the green tapestry of plants, the world of colors—red, green and blue—and the sounds that are out there. What are these fleeting formations that modern-day physics, physiology and psychology regard only as subjective? They are the elements from which the worlds of the future create themselves. Red is not engendered by matter in the eye or the brain, red is the first, semblance-like, seed of future worlds.

If you know this, you will also want to know something about what will correspond in these future worlds to the corpse-like element. It will not be what we found earlier in our physics and chemistry, it will be the corpse of the future. We shall recognize what will be the corpse of the future, the future element of death, if we discover it already today in the higher organization of man, where astral body and ego are active. By experiencing the final condition there in reference to the initial one, we at last gain a proper comprehension of the nervous system and the brain insofar as they are dead, not alive. In a certain sense, they can be more dead than a corpse, inasmuch as they transcend the absolute point of death—especially in case of the nervous system—and become "more dead than dead." But this very fact makes the nervous system and the brain bearers of the so-called spiritual element—because the dead element dwells in them, the final state not yet even reached by outer nature—because they even surpass this final state.

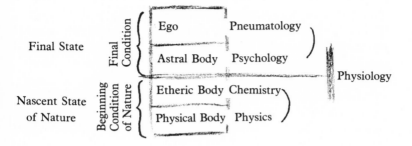

In order to find psychology and pneumatology in the outer world, we shall have to discover how the inanimate, the dead, dwells in the human organism; namely, in the head organization and in part of the rhythmic organization, mainly that of breathing. We must look at our head and say of it that it is constantly dying. If it were alive, the growing, sprouting living matter could not think. But because it gives up life and constantly dies, the soul-spiritual thoughts, endowed with being, have the opportunity to spread out over what is dead as new living, radiant semblance.

You see, here lie the great tasks that, by means of the historical manner of observation, result quite simply from natural science. If we don't take hold of them, we move like ghosts through the present development of science, and not with the consciousness that an epoch that has begun must find a way to continue. You can imagine that much of this is contained implicitly in what science has discovered. Scientific literature offers such indications everywhere. But people cannot yet distinguish clearly; they like what is chaotic. They don't care clearly to contemplate physics and chemistry on one hand, and psychology and pneumatology on the other, because then they would have to consider seriously the inner and outer aspects. They prefer to vacillate in the murky waters between physics and chemistry. Due to this, a bastard science has arisen that has become the darling of natural research and even philosophy; namely, physiology. As soon as the real facts are discovered, physiology will fall apart into psychology on the one hand—a psychology that is also a perception of the world—and on the other, into chemistry, meaning a chemistry that is also a knowledge of man.

When these two are attained, this in-between science, physiology, will vanish. Because today you have a morass in which you can find everything, and because by juggling a bit to the left or the right, it is possible to find a bit of a soul or a corporeal element, people do quite well. The physiology of today is what above all must disappear as the last remnant of former conceptions that have become muddled. The reason physiological concepts are so abstruse is that they contain soul and corporeal elements that are no longer distinguished. It quite suits people that they don't have to distinguish, thus they can play around with words and even juggle the facts. One who aims for clear insight must realize that physiology amounts in the end to fibbing with words and facts.

Until we admit this, we can't take the history of natural science seriously. Science does not proceed only from undetermined past ages to our time, it continues on from the present. History can only be understood, if one comprehends the further course of things, not in a superstitious, prophetic sense but by beginning now to do the right thing. And infinitely much needs to be set right, particularly in the domain of science. Natural science has grown tall; it is like a nice teenager, who at the moment is going through his years of unpolished adolescence, and whose guidance must be continued so that he will become mature. Science will mature, if murky areas like physiology disappear, and physics and pneumatology arise again in the way outlined above. They will come into being, if the anthroposophical way of thinking is applied in earnest to science. This will be the case, when people feel that they are learning something, when somebody speaks to them of a real physics, a real chemistry, a real psychology and pneumatology; when they no longer have the urge to comprehend everything concerning the world and the human being through bastardized chaotic sciences like physiology. Then, the development of human knowledge will once again stand on a sound basis.

Naturally, therapy is particularly affected and suffers under present-day physiology. You can well imagine this, because it works with all manner of things that elude one's grasp, when one begins to think clearly.

We cannot confront the great challenges of our time with a few anthroposophical catchwords and phrases. It also does not suffice to dabble with physiology on the borderline between psychology and chemistry. The only way to proceed is to apply the methods of spiritual-scientific anthroposophy to physics and chemistry. If you are lazy—forgive me for this harsh expression, I don't mean it in such a radical sense in this case—you say: These matters can only be correctly judged, if one is clairvoyant. Therefore I will wait until I am clairvoyant. I won't venture to criticize physics and chemistry or even physiology.

My dear friends, you need not have insights that surpass ordinary perception in order to know that a corpse is dead and that it must have originated in life. Neither do you need to be clairvoyant in order to analyze properly the true facts of today's physics and chemistry, and to refer them back to their underlying living element, once your attention is directed to the fact that this living element is to be found by studying the "lower man." There you will have the supplement you need for chemistry and physics. Make the attempt, for once, really to study the mechanism of human movement.[86] Instead of constantly drawing axis of coordinates and putting the movements into them apart from man; instead of multiplying differential coefficients and integrals, make a serious attempt to study the mechanics of movement in man. As they were once experienced from within, so do you now study them from without. Then you will have what you need, to add to your outer observation of nature, in physics and chemistry.

In outer nature, those who proclaim atomism will always put you in the wrong. They even work themselves up to the very spiritual statement that when one speaks about matter in the sense of a modern physicist, matter is no longer material. The physicists themselves are saying it;[87] our very opponents are saying it. In this case they are right, and if we in our replies to them stop short at the half-truths—that is to say, at the final conditions of being—we shall never be equal to that which issues from them.

Here lie the tasks of the specialists, here lie the tasks of those who have the requisite preliminary training, in one or another branch of science.

Then we shall not establish a physicized or chemicized Anthroposophy, but a true anthroposophical chemistry, anthroposophical physics. Then we shall not establish a new medicine as a mere variation on the old, but a true anthroposophical medicine.

The tasks are at hand. They are outlined in all directions. Just as the simple heart can receive the observations that are scattered everywhere in our lectures or lecture cycles, and that give spiritual sustenance, so too the need is to take up on every hand the hints that can lead us to the much-needed progress in the several domains of science.

In the future, it will not suffice if man and nature do not again become one. What physics and chemistry study in nature as the final state of being, must be supplemented by the state of being in "lower man" belonging to the realm of physics and chemistry—in man who is dependent on the physical and etheric bodies. It is important that this be sought. It is not important to single out as essential the valences of the structural formulas or the periodic law in chemistry, because these are but schemata. While they are quite useful as tools for counting and calculations, what matters is the following realization. If the chemical processes are externally observed, the chemical laws are not within them. They are contained in the origin of chemical processes. Hence, they are found only, if, with diligent effort, one tries to seek in the human being for the processes that occur in his circulation, in the activity of his fluids, through the actions of the etheric body. The explanation of the chemical processes in nature lies in the processes of the etheric body. These in turn are represented in the play of fluids in the human organism and are accessible to precise study.

Anthroposophy poses a serious challenge in this direction. This is why we have founded research institutes[88] in which serious, intensive work must begin. Then the methods gained from anthroposophy can be properly nurtured. This is also the main point of our medical therapy; namely, that the old, confused physiology finally be replaced with a real chemistry and psychology. Without this one can never assert anything about the processes of illness and healing in human nature, because every course of illness is simply an abnormal psychological process, and each healing process is an abnormal chemical process. Only to the extent that we know how to influence the chemical process of healing and how to grasp the psychological course of illness will we attain to genuine pathology and therapy. This will emerge from the anthroposophical manner of observation. If one does not want to recognize this potential in anthroposophy, then one only wants something a bit out of the ordinary and is unwilling to get to work in earnest. Actually, everything that I have sketched here is only a description of how the work should proceed, because a genuine psychology and chemistry come into being through work. All the prerequisites for this work already exist, because very many facts can be found in scientific literature that researchers have accidentally discovered but don't understand. Those of us who work in the spirit of anthroposophy should take up these facts and contribute something to their full comprehension. Take as an example what I emphasized yesterday[89] in speaking to a smaller group of people. The essential point about the spleen is that it is really an excretory organ. The spleen itself is in turn an excretion of the functions in the etheric body. Countless facts are available in medical literature that need only be utilized—and that is the point: they should be *utilized*—then the facts will be brought together and what is needed will result.

A single person might accomplish this if a human life spanned six hundred years. But by that time, other tasks would confront him and his accomplishments would long since be outmoded. These things must be attained through cooperation, through people working together. So this is the second task—we must see to it that this becomes possible. I believe that these tasks of the Anthroposophical Society will emerge most clearly and urgently from a truly realistic study of the history of natural science in recent times.

This history shows us at every turn that something great and wonderful has arisen through modern science. In earlier times, the truly inanimate dead aspects could never be discerned, hence, nothing could be made of them. In those times inward semblance could never really be observed; therefore, it couldn't be brought to life by human effort, and hence, one couldn't arrive at freedom. Today, we confront a grandiose world, which became possible only because natural science studies the dead aspects. This is the world of technology. Its special character can be discerned from the fact that the word "technique" is taken from the Greek. There, it still signifies "art," implying that art reveals, where technology still contains spirit. Today, technology only utilizes spirit in the sense of the abstract, spirit-devoid thoughts. Technology could be achieved only by attaining a proper knowledge of what is dead. Once in the course of humanity's evolution it was necessary to concentrate upon the dead; it thus entered into the realm of technology. Today, man stands in the midst of this realm of technology that surrounds him on all sides. He looks out on it and realizes that here at last is a sphere in which there is no spirit in the proper sense. In regard to the spiritual element, it is important that in all areas of technology human beings experience this inner feeling, almost akin to one of pain over the death of a person. If feeling and sensation can be developed in knowledge, then such a feeling will arise, somewhat like the sensation one experiences, when a person is dying and one sees the living organism turn into a corpse. Alongside the abstract indifferent cold knowledge, such a feeling will arise through the true realization that technology is the processing of the inanimate, the dead. This feeling will become the most powerful impetus to seek the spirit in new directions.

I could well imagine the following view of the future: Man looks out over the chimneys, the factories, the telephones—everything that technology has produced in wondrous ways in the most recent times. He stands atop this purely mechanical world, the grave of all things spiritual, and he calls out longingly into the universe—and his yearning will be fulfilled. Just as the dead stone yields the living fiery spark if handled correctly, so from our dead technology will emerge the living spirit, if human beings have the right feelings about what technology is.

On the other hand, one need only understand clearly what pure thinking is; namely the semblance from which can be brought forth the most powerful moral impulses—those individual moral impulses that I have described in my *Philosophy of Freedom*. Then, in a new way, man will face the feeling that was once confronted by Nicholas Cusanus and Meister Eckhart. They said: When I lift myself beyond everything that I am ordinarily accustomed to observe, I come to "nothingness" with all that I have learned. But in this "nothingness" there aises for me the "I." If man really penetrates to pure thinking, then he finds in it the nothingness that turns into the I and from which emerges the whole wealth of ethical actions that will create new worlds. I can imagine a person who first lets all knowledge of the present, as inaugurated by natural science, impress itself on him and then (centuries after Meister Eckhart and Nicholas Cusanus) turns his gaze inward and with today's mode of thinking arrives at the nothingness of his inner life. In it, he discovers that the spirit really speaks to him. I can imagine that these two images merge. On the one hand, man goes to the place where barren technology has left the spirit behind. There he calls out into cosmic expanses for the spirit. On the other hand, he stops, thinks and looks within himself. And here, out of his inner being, he receives the divine answer to the call he sent out into the distances of the universe.

When we learn, through a new, anthroposophically imbued natural science, to let the calls of infinite longing for the spirit, sent out into the world, resound in our inner being, then this will be the right starting point. Here, through an "anthroposophized" inner perception, we will find the answer to the yearning call for the spirit, desperately sounded out into the universe.

I did not want to describe the development of natural science in recent times in a merely documentary fashion. Rather, I wanted to show you the standpoint of a human being, who comprehends this natural-scientific development and, in a difficult moment of humanity's evolution, knows the right things to say to himself in regard to the progress of mankind.

NOTES

1. Rudolf Steiner, *Mysticism at the Dawn of the Modern Age* (Blauvelt, NY: Steinerbooks, 1960) (formerly published as *Eleven European Mysteries*).

2. These include the three natural scientific courses held in Stuttgart: *First Scientific Lecture Course: Light Course* (Forest Row, England: Steiner Schools Fellowship, 1977); *Warmth Course* (Spring Valley, NY: Mercury Press, 1981); and *Das Verhältnis der verschiedenen naturwissenschaftlichen Gebiete zur Astronomie.* (Dornach, Switzerland: Rudolf Steiner Verlag). The relationship between natural science and spiritual science is dealt with in *The Boundaries of Natural Science* (Spring Valley, NY, Anthroposophic Press, 1983).

3. Nicholas Cusanus (Nicholas of Cusa), 1401–1464. Lawyer, churchman, philosopher, mathematician. Ordained priest between 1436–1440, Cardinal 1448. Bishop of Brixen, 1450. cf. chapter on Cusanus in *Mysticism at the Dawn of the Modern Age.*

4. Nicholas Cusanus was made Cardinal and named Bishop of Brixen in rapid succession. Though a stranger to Brixen he was named Bishop there directly by the Pope. This led to a protracted conflict with his diocese, during which the latter gathered behind the Duke of Tirol. Cusa was ambushed by the Duke, imprisoned, and forced into accepting a demeaning agreement. The Duke was excommunicated by the Pope and attacked by the Swiss Confederation. However, he was supported by German Counts and remained intransigent. Cusa died before the Emperor could resolve the conflict. The battles around him did not rob Cusa of his peace of mind, and he developed his philosophic, mathematical and theological insights, writing fifteen of his works during the time in Brixen.

5. *Brethren of Common Life* (also *of Good Will*): Founded by Gerhart Groote around 1376. Brother-houses in Holland, Northern Germany, Italy and Portugal. Brought into the Catholic Church in the Fifteenth Century. Their schools taught under the strict observance of dogma.

6. *Council of Basel*: 1431–1449. Called by Pope Martin V on July 23, 1431, the year of his death. This was the last of four reformatory councils with the aim of ending the division in the Church. There came a new rift in the Church.

7. In 1437. This summarizes a long process: Cusanus entered the Council 1432 with the task from the Archdiocese of Trier to defend their Archbishop, whom they had chosen against the will of the Pope. Through the treatise *De Concordantia Catholica* (On Catholic Unity) which he distributed among the Council and which contained an exceptional survey of the decisions of the Councils and Decrees of the Church, he offered the advice welcome by the majority that the Common Council was beyond the Pope. Thus, he immediately became an important figure in the Council.

Later, the Council majority and the history writings accused Cusanus of having changed his conviction. But Cusanus' deep understanding was ignored, which was rooted in his attitude and which comes to expression in the following words: "When a decision is made unanimously, then one can believe that it came from the Holy Spirit. It lies not in men's power to meet somewhere, and although they are so different from each other, they are able to come to a harmonious decision. It is God's work. (From J.M. Duex, *Der Deutsche Cardinal Nicolaus Von Cusa*, Regensburg 1874, Bd. 2, s. 262, which has translated some of the most important of the *De Concordantia Catholica*. Cusanus must have experienced at the Council that his description of the meaning of a Council was not taken with interest, and he must have faced a decision that is mentioned in the lecture.

8. Pope Eugene 4th was put down and Duke Amadeus of Savoy was set up as Pope Felix 5th in 1439. His resignation in 1449 caused the disbandment of the Council.

9. From 1439–1448 Cusanus acted on the order of the Pope as "Hercules of the Eugenians" as an opponent called him. He went to worldly and churchly princes as well as to the "Reichstag," and he tried to overcome the neutrality of the Germans about the split of churches, with complete success.

10. At the meetings of the princes, 1454, in Nurenberg, Regensburg, and Frankfurt after the invasion of Constantinople by the Turkish, Cusanus tried to motivate the princes to a crusade. After J. Hunnyadis' victory over the Turkish Army in front of Belgrad in 1456 Cusanus organized, at the same day he received the message, a festival of thanksgiving, and he spoke the following words: "Because the lower man can only enjoy life animal-like and physical, Satan who wants to destroy the Gospels in a fine way, intended the appearance of Muhammed who knows the Gospel and the Bible, to let him give the Gospel

and Bible an animal-like, sensual meaning. In this way Satan taught Muhammed knowledge to let go forth the head of Maliguity, the son of Ruin, and to be an enemy of the cross of Christ. (From a sermon, *"Landans Invocalo Dominum,"* partly translated by J.M. Duex A.A.O.S. 165). Further sermons against the Turks are known from October 28, and November 5 of the same year. (E. Varisteenberge, *Le Cardinal Nicolas De Cues*, Paris 1920, S. 231 F, and index of sermons s. 480), but this sermon seems to be available only in Latin.

Cusanus himself announced his appointment as Cardinal with a short autobiographical note in which is written: Nicolas was made Cardinal secretly by Pope Eugene (*Hist. Jahrbuch der Goerrers Gesellschaft* 16.S.549).

11. *De Pace Fidei* (*On the Peace of the Faiths*), written in September 1453. "The horrible days of Constantinople . . . had caused a deep feeling of sadness in the breast of a man who once had wandered through this region, and caused him to sink into deep contemplation, and he had a vision. In this sublime state, he particularly thinks about the differences of the religions of the world, and the possibility of their harmony. This harmony is, in his opinion, a basic condition for religious peace." (Introduction to *De Pace Fidei*: Nach Duex A.A.O. S.405).

12. Cusanus left Basel in May 1437 together with other representatives of the minority and travelled for the minority with the legation of the Pope to Constantinople to accompany the Greek Emperor and the heads of the Eastern Church to the Union Council in Ferrara. They arrived in February 1438 in Italy.

13. *De Docta Ignorantia* (*The Learned Ignorance*). Three books finished in February 1440.

14. See Rudolf Steiner, *The Redemption of Thinking*. (Spring Valley, NY: Anthroposophic Press, 1983).

15. Meister Eckhart: Hochheim by Gotha about 1260—before 1328, Cologne. Dominican, schoolmaster, German mystic. Preached in leading posts in orders and churches; taught in Paris, Strasbourg, Cologne. Main work: *Opus Tripartius*. Based on Scholasticism and writings of Dionysius the Areopagite. Copies of his sermons partly went around without his control. Meister Eckhart died, accused as heretic, during the trial. See chapter, "Master Eckhart," in Rudolf Steiner's *Mysticism at the Dawn of the Modern Age*.

16. These lines cannot be made clear and simple because the German text plays at length on the words *Nicht* and *Ich*.

135

17. Thomas Aquinas: Castle Roccasecca in the Neopolitan region, about 1225–1274 Cloister Fossanuova. Dominican, scholar, churchman. In Cologne, student and friend of Albertus Magnus. Advocated the spiritual reality of general concepts. He directed the theological school in Rome from 1261–1267. There the studies of the Dominican; from 1268 onwards he is teaching in Naples and France. See Rudolf Steiner, *The Redemption of Thinking* and *The Riddles of Philosophy* (Spring Valley, NY: Anthroposophic Press, 1973).

18. Nicholas Copernicus: Thorn 1473–1543 Frauenburg. Humanist, mathematician, astronomer, physician, lawyer. No publications during his life, with the exception of a translation. Finished his work on the heliocentric planetary system around 1507. Copernicus was already on his deathbed when his *De Revolutionibus Orbium Coelestium* was published. He dedicated it to Pope Paul III. His friend and publisher introduced it as a purely hypothetical, special scientific method of calculation. It thus slipped past the censor, until the third edition was banned in 1616/17. Not until 1822 was it accepted by the Catholic Church. cf. Rudolf Steiner, *The Spiritual Guidance of Man.* (Spring Valley, NY: Anthroposophic Press, 1983).

19. Post-Atlantean Age: cf. Rudolf Steiner, *An Outline of Occult Science* (Spring Valley, NY: Anthroposophic Press, 1984).

20. A literal translation of the transcript would read: "As body; and as body, as an image of the spirit."

21. I listen to the silent universe: cf. Rudolf Steiner, *Truth-Wrought Words.* (Spring Valley, NY: Anthroposophic Press, 1979).

22. Democritus: c. 460–360 B.C. From his numerous writings about philosophy, mathematics, physics, medicine, psychology, and technology, only some fragments and an index remain. The remark mentioned is a report from Aristotle, *Metaphysics* I:4: "That is why they (Leucippus and Democritus) say that the non-existent exists just as much as the existent, just as emptiness is just as good as fullness, and they posit these as material causes."

23. Francis Bacon: (also Francis Bacon of Verulam), London 1561–1626 Highgate. Lawyer, doctor, politician, diplomat, essayist, philosopher and humanist. The leading English government liberal, successful during 1603–1621. In these years his main work was developed. The philosophy of his age he found stuck in hopeless experiments to solve insoluable problems with Aristotelian logic. The only source of sure knowledge and abilities for him was natural science. He saw a renewal

of the spiritual and economic life in this science. Principal works: *Novum Organum* (an inductive logic contradicting that of Aristotle (thd old *Organum*); *De Dignitate et Augmentis Scientiarum: a Critical Encyclopedia of all Science*; *Sylva Sylvarum: Preliminary Announcement of Procedure and Method* (this remained in preparation). His literary success was astonishing, and it greatly furthered the materialistic world view. cf. *The Riddles of Philosophy.*

24. Spinoza, Benedictus: Amsterdam 1632–1677. The Hague. Philosopher, mathematician, had Humanistic and Talmudistic training. By vocation, optician and politician. His main work *Ethics* with the characteristic full title *Ethica Ordine Geometrica Demonstrata* (*Ethic Represented by Geometric Method*) could only be published by his friends after his death. See *Mysticism at the Dawn of the Modern Age* and *The Riddles of Philosophy.*

25. René Descartes: Lat., Renatus Cartenius, Le Haye (Tourraine) 1598 –1650 Stockholm. Mathematician, physicist, philosopher. Educated by the Jesuits in La Fleche, he first became a soldier and was part of some campaigns but turned away from outer life to enter into the loneliness of a striver for knowledge, living first in Paris and then for a long time in Holland. He died in Stockholm, having been called there by Queen Christine. For him, doubt of tradition, but also of all sense perception, was the starting point of his philosophy and he found in self-consciousness the security of all being (*"Cogito ergo Sum"*). He developed the method of analytical geometry and gave an explanation of the rainbow. Main works: *Essays*, 1637, in it *"Discours de la Methode and Dioptiric,"* *"Meditationes de Prima Philosophia,"* 1641; *"Passions de L'Ame,"* 1650. See *The Riddles of Philosophy.*

26. Non-Euclidian geometry is a prime example of "the self-contained inner ability to think." C. Friedrich Gauss (1777–1855) discovered first that one can think more than only a geometric system. Because nobody understood this, he decided not to publish his results and to withdraw from the fruitless quarrel. Independently of Gauss in 1828 N.I. Lobatschewskij and in 1829 J. Boljai first published their solutions to the same problem. Rudolf Steiner often spoke about the meaning of this achievement, as in *Wege und Ziele des geistigen Menschen* in the lecture *"Der Heutige Stand der Philosophie und Wissenschaft,"* (Dornach, Switzerland: Rudolf Steiner Verlag, 1973; GA Bibl. Nr. 125). See also: Georg Unger, *Physic am Scheidewege* (Dornach: 1948), pages 19–28, and *Vom Bielden Physikalischer Begriffe*, vol. 3 (Stuttgart: 1967), pages 31–32 and 193–194.

27. Johannes Tauler: About 1300–1361 Strasbourg. Preacher and pastor, Dominican, mystic, student of Meister Eckhart. Sermons and writings in German by W. Lehmann, 1923; see also *Mysticism at the Dawn of the Modern Age*, the chapter "Friendship with God."

28. Rudolf Steiner, *The Case for Anthroposophy* (London: Rudolf Steiner Press, 1970).

29. In a reply to two lectures, which Walter Johannes Stein and Eugen Kolisko gave to defend two articles on "Anthroposophy as Science" in the Goettingen newspaper, Hugo Fuchs, Professor of Anatomy, spoke sarcastically of a human being with head, breast, and belly system. (From a report of the newspaper *Die Dreigliederung des Sozialen Organismus*, August 1920, No. 5).

30. From Goethe's *Faust*, Part I, the scene in the Student Room with Faust and Mephisto. See Rudolf Steiner, *The Occult Significance of the Blood* (London: Rudolf Steiner Press, 1967).

31. William Harvey, 1578–1658, physiologist, Professor of Anatomy, London, discoverer of the main bloodstream: *De Motu Cordis et Sanguinis* (1628).

32. Giordano Bruno: Nola 1548–1600 Rome. Dominican, 1563–1576, a great traveler. Main works developed at the English court at the time of Elizabeth I. After he returned to Italy he was imprisoned because of heretical teachings, and was burned in Rome after 8 years in prison. See *The Riddles of Philosophy*, and *The Spiritual Guidance of Man*, by Rudolf Steiner.

33. Isaac Newton, Sir: Woolsthorpe, Lincolnshire 1642–1727 Kensington, London. Born as a dwarf-like child. Grew up on a farm and went to village and small town schools until 1661. After he was accepted at the University he was of medium talent until his "flaming" as a genius physicist, astronomer, mathematician 1663–1664. Professor in Cambridge 1669–1701, member of the Royal Society London 1662 and from 1703 until his death, its President. Main work: *Law of Gravitation, Mathematically Adapted to the Law of Motion from Kepler*, developed 1666, published 1687 in *Philosophiae Naturalis Principa Mathematica*. The idea of an infinitesimal mathematics came from Newton in 1663; three years later he had developed his differential mathematics. His *Optics*, 1704, put forth the division of light in color as well as emission theory.

   Later Newton lost all interest in physics, mathematics, and also in the destiny and consequences of his works. He turned towards chemical and alchemical experiments and studies of their old traditions. In

138

his old age he was interested in religious-speculative studies. Before his death he compared his life with a day, in which a child is playing with sand and mussels and is not aware anymore of the cosmos at his back. Literature: J.W.N. Sullivan, *Isaac Newton 1642–1727* (London 1938).

34. In Newton's second edition of his *Philosophiae Naturalis Principa Mathematica* of 1713 the definition is "But I do not define, because it is well known to all of us."

35. George Berkeley: Desert Castle, Thomastown, Ireland 1685–1753 Oxford. English philosopher and Anglican missionary, Bishop from 1734. Main works: *Treatise Concerning the Principle of Human Knowledge*, 1710; *Alciphron*, about ethics and free thinkers, 1732; *Siris*, concerning metaphysical questions. See: *The Riddles of Philosophy*.

    Berkeley said: "One has to do it in such a way": e.g., as in Paragraph 113 of *Principles of Human Knowledge*. In the writing *De Motu* (*From Motion*) is written in Paragraph 43: "Motion, even though perceived clearly by the senses, was darkened, but not because of its own being, but far more through commentaries by learned philosophers."

36. In the work *Optice* by Newton, which is the Latin translation of his *Optics* (1704), published by Samuel Clarke in 1706 and approved with additions made by Newton, the formula appears only at the end of the book at the so-called 28th Problem: "If these questions are answered in the right way, could we then not ascertain the phenomenon that there is a being, unbodily, intelligent, which can perceive the endless universe as it were with its sense organs, and which seems to look into the innermost and is surrounding it with its all-embracing presence, while that in us that is usually feeling and thinking are only handed-down pictures in which we then perceive and observe our organs?" This thought seems not only to be Newton's, but was also presented in a similar way by Henry More, the Platonist from Cambridge who was a friend of Newton.

37. For his polemic concerning Newton's color theory, see Rudolf Steiner, *Goethe the Scientist* (New York: Anthroposophic Press, 1950), especially the Introduction, "Goethe, Newton and the Physicists"; see also the forthcoming book, Heinrich O. Proskauer, *The Rediscovery of Color* (Spring Valley, NY: Anthroposophic Press).

38. Leibnitz: Leipzig 1646–1716 Hanover. Philologist, mathematician, physicist, lawyer, statesman, priest. Mostly living at princely courts, traveling a lot. Discoverer of the Infinitesimal Calculus 1686, independently of Newton.

39. In his writing *The Analyst* (*The Analyst*, 1734, included in the book *Writings about the Origin of Mathematics and Physics*) the Table of Contents is in the form of 50 theses. No. 7, for example, is as follows: "Objections against the Secrets of Belief Which are Made Unfairly by Those who Admit Them in Science;" or No. 13: "The Rule for the Flux of Potency is Achieved through Unfair Reasoning;" and No. 22: "With the Help of a Double Mistake Analysts Come to their Truth, but not to Science, in which They do not even Know How They Came to Their Own Conclusions." From the *Polemic Dispute*, which follows *The Analyst*, an example is: "No big name on this earth will ever cause me to take unclear things for clear ones. They think of one as if it were a crime to think one could see further than Sir Isaac Newton, even above him. I am convinced though that they speak for the feelings of many others. But there are also some . . . who think and feel it unfair to copy some great man's shortcomings, and who see no crime in wanting to see further than Sir Isaac Newton, but further than the whole of mankind."

40. Prepared by Mach and Lorentz, developed by Einstein, *Special Theory of Relativity* 1905, *Common Theory of Relativity* 1916. Made it necessary to revise Newton's Mechanics with the help of non-Euclydean Geometry. See also *The Riddles of Philosophy* and Georg Unger, *Von Bielden Physicalischer Begriffe*, Part 3 (Stuttgart: 1967), pages 100–122.

41. Lessing, Gottfried Ephraim: Kamenz/Lausitz 1729–1781 Brunswick. Dramatist, essayist, critic. Opens a new epoch in German literature and art. His last writing, "The Education of the Human Race," (1780) finds it necessary to postulate reincarnation for the sake of the development of the human race. See *The Riddles of Philosophy*.

42. The reason was a controversy in the magazine *Die Drei* of 1921–1922, pages 1107 and 1114, as well as in the following years publication (see pages 172–330 about the reality of atoms). See Rudolf Steiner's *First Scientific Lecture-Course: Light Course* (Forest Row, England: Steiner Schools Fellowship, 1977).

43. John Locke: Wrington by Bristol 1632–1704 Oates, Essex. Theologian, philosopher, and physician. Because raised Protestant and Puritan, he was persecuted in England and had to flee to Holland until after the English Revolution of 1689. From 1675–1689 Locke worked with many interruptions at his main work, *An Essay Concerning Human Understanding*, 1690. Originally he had planned a critical presentation of the already recognized teaching of primary and secondary sense characteristics, but then it grew to a perception theory

or world view. His *Essay* was published 4 times in his lifetime. See *The Riddles of Philosophy, The Philosophy of Freedom*, trans. Michael Wilson (Spring Valley, NY: Anthroposophic Press, 1964) and "Cardinal Nicolas of Cusa" in *Mysticism at the Dawn of the Modern Age.*

44. Richard Wahle: 1857–1935, Vienna, Professor of Philosophy. Only valued perceptions, imaginations, and feelings, but rejected all philosophy hitherto written as theories of cognition. The "Ego" is for him "a summary of surface-like, physiologically accompanied pieces of consciousness, which are brought into being by invisible forces." Some writings: *The Whole of Philosophy and Its End*, 1894; *About the Mechanism of the Spiritual Life*, 1906; *The Tragic Comedy of Wisdom*, 1915; *Development of Characters*, 1928; *Basics of a New Psychiatry*, 1931.

45. See Rudolf Steiner, *The Philosophy of Freedom*, Chapter 4.

46. Immanuel Kant: 1724–1804. Lived in Koenigsberg, which he seldom left. Philosopher, scientist, mathematician. Professor in Koenigsberg 1770–1794. *Critique of Pure Reason*, 1781. Its popular edition *Dissertation on Any Future Metaphysics*, 1783, his ethic *Critique of Practical Reason*, 1788, aesthetic and natural theology is handled in *Critique of Judgment*, 1790. He wrote the first mechanical cosmology 1755. It was taken up and changed by Laplace (1796) and known as the Kant-LaPlace Theory. Rudolf Steiner's exposition on Kant's theory is found in *Truth and Knowledge, The Philosophy of Freedom*, and *An Autobiography*, ed. Paul M. Allen, 2nd ed. (Blauvelt, NY: Steinerbooks, Garber Communications, 1980).

E.g. in *Critique of Pure Reason*, "Transcendental Aesthetic, Common Remarks": "We wanted to say that all our opinions are nothing but the conception of the appearance; that the things we look at are not actually what we take them for, nor is their relation constituted as they appear to us, and that if we would suspend our subject or even our subjective constitution of our senses as a whole, the whole constitution, all relationships of objects in space and time, even time and space itself would disappear. They would only exist in us, not as phenomena in themselves."

47. August Weismann, Frankfurt A.M. 1834–1914 Freiburg. Biologist, genetic scientist. Theory of polarity between cells (soma) and seed plasma. Determinants as heredity carriers. Writing: *Studies on the Descent Theory.*

48. Goethe's recital from *Faust* I, Act 1, Scene 2, "Night," Gothic Room, Wagner and Faust:

"My friend, the time of past
Is a book with seven seals.
What you call the Spirit of Time
Is fundamentally the Gentleman's own spirit,
In which the times reflect themselves."

49. Henry Poincaré: Nancy 1854-1912 Paris. Author of the popular philosophical writings *Science and Hypothesis* (1902), *The Value of Science* (1905), *Science and Method* (1909), and *Last Thoughts* (1912). The lecture in question was held by Poincaré shortly before his death in a lecture cycle *Conferences de Foi et de Vie* printed in *Le Materialisme Actuel* with M.M. Bergson, H. Poincaré, Ch. Gide, Ch. Wagner, Firm Roz, De Witt-Guizotfriedel, Gaston Rion. (Paris: 1918), page 53.

50. Mathias Jakob Schleiden: Hamburg 1804-1881 Frankfurt A.M. Lawyer, physician, and, mainly, biologist. Developed a cell formation theory in *Contributions to Phytogenesis* (1838).

51. Theodor Schwann: Neuss 1810-1882 Cologne, biologist. Founded the cell theory with his *Microscopic Examinations of the Harmony in Structure and Growth of Animals and Plants* (1839).

52. In the night from New Year's Eve to New Year's Day 1922/23 the Goetheanum burned down. It was built in ten years, with the help of various artists from many countries. This primarily wooden building, in which each surface and corner was formed artistically (see Steiner, *Ways to a New Style in Architecture* [London: Anthroposophical Publishing Co. and New York: Anthroposophic Press, 1927]) had been designed in all details by Rudolf Steiner who also managed the construction work through all these years. From the first of January on, the activities had to be transferred into the so-called "Schreinerei," a building that was used during the construction of the Goetheanum. For the work itself, Rudolf Steiner did not allow any interruption; the afternoon after the fire, the "Three Kings Play" was performed, as was written on the invitations of the ongoing course (see *Christmas Plays from Oberufer*, trans. A.C. Harwood [London: Rudolf Steiner Press, 1973]). Rudolf Steiner introduced it with a short address, in which he spoke the following words: "great suffering knows how to keep silent about what it is feeling . . . The building that was created in ten years through the love and compassion of innumerable friends of the movement was destroyed in one night. But just today the silent suffering experiences what our friends have put in this work. Since we feel that everything we do in our movement is necessary in our present civilization, we will want to continue whatever we can in the

142

given frame, and therefore even in this hour as the flames outside still burn and rise, although such suffering is present, still perform this play which we promised our participants in connection with our course, and which these participants expect. I also will hold the lecture I offered, here in the "Schreinerei" this evening at 8:00 P.M. (printed in *Ansprachen zu den Weihnachtsspielen aus Altem Volkstum* [Dornach: Rudolf Steiner Verlag, 1974], GA Bibl. Nr. 274). The beginning of the course's lecture was then devoted to the fire, which is printed in *The Younger Generation* (Spring Valley, NY: Anthroposophic Press, 1984).

53. One can find the basic reality explained in the chapter "Sleep and Death" in *An Outline of Occult Science* and in *Knowledge of the Higher Worlds and Its Attainment* (Spring Valley, NY: Anthroposophic Press, 1983).

54. Paracelsus, Theophrastus von Hohenheim: Einsiedeln, Kanton Schwytz 1493-1541 Salzburg, Md. Ferrara, Professor in Basel. Accomplished physician, scientist, and philosopher. Wrote about chemistry, medical science, biology, astronomy, astrology, alchemy, and theology. The myths about Paracelsus as goldmaker, magician, or charlatan were made up after his death and distorted the picture of his character. Most complete work published by Karl Sudhoff (fourteen volumes). See *The Riddles of Philosophy*.

55. Helmont, Johann Baptist van: Brussels 1577-1644. Physician and iatrochemist. He managed the differentiation and separation of gases (hydrogen, carbon). He coined the name "gas" for the airy state.

56. See Steiner, *Goethe the Scientist.* Especially see Chapters 7, 8, 9, 10, 11, 12, 15.

57. Lecture of April 8, 1911, at the 9th International Philosophical Congress, "The Psychological Foundations of Anthroposophy," in Rudolf Steiner, *Esoteric Development*, Spring Valley, NY: 1982), pp. 25-55.

58. See Steiner, *Goethe the Scientist.*

59. Rudolf Steiner, *Theosophy* (Spring Valley, NY: Anthroposophic Press, 1971), pp. 1-39.

60. Rudolf Steiner, *Man and the World of Stars: The Spiritual Communion of Mankind* (New York: Anthroposophic Press, 1963), pp. 141-172.

61. Galileo Galilei: Pisa 1564-1642 Arcetri by Florence. Discovered isochromism in pendulum, hydrostatic scales, laws of free fall, law of inertia. Numerous astronomical inventions with self-constructed

telescope. An Inquisition trial resulted in a banning of the Copernican world system. See *The Riddles of Philosophy, The Spiritual Guidance of Man*, and Laurenz Muellner's speech, "*Die Bedeutung Galilei's fuer die Philosophie,*" Vienna 1894. (Reprinted in *Anthroposophie*, 1933/34:29).

His *Sermons de Motu Gravium (About the Effects of Gravity)* contain the results of his investigations in Pisa. They first only circulated in manuscript copies; first edition: 1854. The final version is in the *Discorsi e Dimenstrazioni Mathematiche Intomo a Due Nuove Scienze*, published 1638 in Leyden. Also see L. Muellner's speech.

62. Such opponents were Bacon, Bruno, Galilei. See *The Riddles of Philosophy* and the speech of L. Muellner, p. 103.

63. Johannes Kepler: Weil der Stadt (Wuerttemberg) 1571-1630 Regensburg. Mathematician, physicist, astronomer, discoverer of the astronomical telescope. Astronomer and mathematician to three emperors. Persecuted as a Protestant. Totally exhausted through his life misery, he died prematurely at the "Reichstag" at Regensburg, where he hoped to secure his subsistence. To calculate his three laws of the motion of the planets he used the observation data of Tycho Brahe, whose follower he was at the court of Prague. On the other hand, the Copernican planetary system was the starting point for the finding of the three laws of the planets. Kepler was the first who tried to interpret the motion of the planetary orbit and moved the center of force to the sun. See *The Spiritual Guidance of Man* and, about the three planetary laws *Das Verhaeltnis der Verschiedenen Naturwissenschaftlichen Gebiete zur Astronomie* (Dornach: Rudolf Steiner Verlag, 1981), GA Bibl. Nr. 323.

64. Galen: Pergamon, Asia Minor 129 A.D.-199 Rome. Physician and philosopher. Studies in Pergamon and travel for study to Corinth, Smyrna, and Alexandria. Personal physician of Emperor Marcus Aurelius. His one hundred and fifty medical texts with fifteen commentaries were the basis for future medicine and pharmacology. One hundred twenty-five texts concerning philosophy, mathematics, and jurisprudence.

65. Rudolf Steiner, *A Road to Self Knowledge: The Threshold of the Spiritual World* (London: Rudolf Steiner Press, 1975), pp. 19-27, 100-106.

66. This is confirmed in chemical textbooks. They speak of chemistry as "a primarily empirical science." In its laws one cannot come to mathematically definite values but to approximate numbers, whose limits are defined in tabular form. Therefore authors of chemical

subject books need to add limiting explanations, such as "usually is valid," or "generally one can say." Chemical laws are mostly derived from physical laws, as for instance in the main theses of thermodynamics. It is thought unscientific to think otherwise than mechanically. Literature: H. Remy, *Lehrbuch der Anorganischen Chemie*, 7th ed., 2 vols. (Leipzig: 1954), Volume I, pages 14–23, 37, 50, 71–73.

67. See Georg Unger, *Vom Bilden Physicalischer Begriffe*, Volume 1, pages 41–49 and 57.

68. See Footnote 40.

69. Steiner, *The Boundaries of Natural Science*, pp. 59–87. Chapters 5 and 6, as well as 7 and 8.

70. Johannes Scotus Erigena: also Eriugena, Ireland 810–877 France. Pre-Scholastic philosopher, theologian with extensive comprehension of language. Came from Britain to France. Led the Emperor's Academy in Paris 845–877. Finished his translation of Dionysius the Areopagite in 858. His main work was *De Divisione Naturae (Division of Nature)*, 867. He taught out of a Platonic comprehension. He stood up for the introduction of the hierarchical order in the worldly administration of the church. See also *Mysticism at the Dawn of the Modern Age*.

71. The copy in Greek originated in the fifteenth century. Dionysius was a member of the Areopag in Athens and a student of the Apostle Paul (Acts 17:34). The setting up of the 3 times 3 hierarchies by Dionysius was adapted as dogma by the Catholic church. His writings in Latin translation were taken up enthusiastically, and were still taken as authentic in the seventeenth century. See *The Riddles of Philosophy, The Redemption of Thinking, Die Ursprungsimpulse der Geisteswissenschaft* (Dornach: Rudolf Steiner Verlag, 1974), GA Bibl. Nr. 96, GA, and Otto Willmann, *Geschichte des Idealismus*, Volume II, paragraph 59.

72. See Steiner, *The Riddles of Philosophy*.

73. Thales of Milet: About 650–560 B.C.

74. Heraclitus of Ephesus: About 550–480 B.C.

75. See also the personalities spoken of in *Mysticism at the Dawn of the Modern Age*.

76. Jacob Boehme: Altseidenberg, Goerlitz 1575–1624 Goerlitz. Mystic. His profession was shoemaker. First writing *Aurora*, 1612. Further writings from 1619 onwards, despite the prohibition. See *Riddles of Philosophy*.

145

77. Iatrochemistry: Name from the Greek "Iatro," physician. Work with homeopathic remedies in continuation of Paracelsus' (1493-1541) method of healing, in the beginning with retention of his opinion about sulfur, mercury, and salt. The Iatrochemical School was established during Paracelsus' last years of life. It degenerated in the middle of the seventeenth century. In its place stepped Robert Boyle's chemistry (1627-1691), for which iatrochemistry had done good preparation. J.B. van Helmont (1577-1644) was one of the main contributors to Iatrochemical literature.

78. Iatromechanics and Iatromathematics. In the sixteenth and seventeenth centuries the proponents of these teachings were nearly all university professors, while iatrochemistry was represented by a union of practicing physicians. But that was true only in the Romantic countries and England. In Italy the main universities were Padua, Pisa, and Rome. There the teachings were rejected on philosophic grounds, because they were based on experience. Germany, where both branches worked hand in hand, was an exception and in a special position.

79. Georg Ernst Stahl: Ansbach 1650-1734, Berlin. Physician and chemist, Professor of Medicine. Exponent of Animism and Vitalism and the hypothesis of the "life forces" in his major work *Theoria Medica Vera*, 1707.

80. Offray de la Mettrie: Malo 1709-1751 Berlin. Physician and writer. Main work is *L'Homme Machine*, published in Leyden 1748.

81. Baron Dietrich von Hollenbach: Heidesheim, Rheinpfalz 1723-1789 Paris. His main work *Systeme de la Nature ou des Lois du Monde Physique et du Monde Moral* appeared 1770 under the pseudonym *Mirabaud*. He only recognized mobile, material atoms, even in regard to thinking, and he based morals on self love.

82. Thomas Hobbes: Malmesbury 1588-1679 Hardwicke. English natural philosopher and humanist. *Opera Philosophica*, 1688. All phenomena in nature and humanity, even the psychological ones, are result of mobility of bodies. The social processes are traced back to mechanical processes. The leading force in this process is the egoism of the single human being. The state which is "crushing everything underfoot," he called "Leviathan" and said: "The natural social condition is the war of all against all."

83. See Drawings, pages 92, 95 and compare with the ones on page 125.

84. See footnote 45.

85. See Rudolf Steiner, *The Karma of Vocation* (Spring Valley, NY: Anthroposophic Press, 1984).

86. Literature: Adolf Fink has studied the mechanism of human movement and the heat produced by muscular work, and published in 1857, 1869, and 1882 *Gesammelte Schriften*, (1903–06 in German).

87. In the beginning of the century Rudolf Steiner pointed to the speech of the philosopher and Prime Minister A.J. Balfour of August 17, 1904, in front of the British Association, immediately after it was held; see Rudolf Steiner, *Lucifer Gnosis* (Dornach: Rudolf Steiner Verlag, 1969), GA Bibl. No. 34, p. 467. Often Steiner also mentioned the lecture of Max Planck of 1910: "*Die Stellung der Neueren Physic zur Mechanischen Weltanshaung*" in Max Planck, *Physikalische Abhandlungen und Vorträge* (Brunswick, Germany, 1958), Vol. 3, pp. 30–46.

88. In 1920 a research institute was founded in Stuttgart for physics and chemistry, with a biological branch through the joint stock company "*Der Kommende Tag.*" A few years later it was transferred to Dornach. The first works from the Institute were published in *Der Kommende Tag: Scientific Research Institute News*. It contains Heft I (1921), "*Milzfunktion und Blaettchen Frage*" von L. Kolisko; Heft II (1923), "*Der Villardsche Versuch*" von Dr. Rer. Nat. R.E. Maier; Heft III (1923): "*Physiologischer und Physicalischer Nachweis Kleinster Entitaeten*" von L. Kolisko. Later works appeared in the volumes *Gaea Sophia, Jahrbuch der Wissenschaftlichen Sektion der Freien Hochschule fuer Geisteswissenschaft am Goetheanum*, Volume I (1926), etc.

89. From this scientific discussion of January 5 no known copy exists.